T0340700

Time to Shine

Scrivener Publishing
100 Cummings Center, Suite 41J
Beverly, MA 01915-6106

Publishers at Scrivener
Martin Scrivener (martin@scrivenerpublishing.com)
Phillip Carmical (pcarmical@scrivenerpublishing.com)

Time to Shine

Applications of Solar Energy Technology

Michael Grupp

With Marlett Balmer, Chris Butters, Jochen Cieslok,
Gerd Schröder
Special guest: Adrien Fainsilber
Thanks to: Joan Beall, Hannelore Bergler, Pierrette Dô,
Mireille Gealageas and Jacques Graff

Scrivener

Co-published by John Wiley & Sons, Inc. Hoboken, New Jersey, and Scrivener Publishing LLC, Salem, Massachusetts.
Published simultaneously in Canada.

Cover design by Kris Hackerott.

Library of Congress Cataloging-in-Publication Data:

ISBN 978-1-118-01621-3

Printed in the United States of America

10 9 8 7 6 5 4 3 2 1

Contents

Summary

Starting from the incoming solar radiation, a subjective overview of the status quo of solar energy use is presented, including the relevant energy transformation processes, solar materials, components, performance and durability requirements. Selected historical, available, and not yet available products are shown, as well as recent and historica examples of solar technology, in particular architecture.

Technical and general development aspects are discussed, including test and monitoring methods, use rate metering, and market introduction experiences.

Outlooks are presented on solar energy in a crowded world, evolving grids, and energy planning aspects.

Reflections in a concentrating solar cooker (Photo: Chris Butters).

About this Book

This book presents an original view of solar energy technology, as well as tools for the understanding of energy-related decisions taken by present and future stakeholders. Particular attention is paid to background phenomena that are not easily found in traditional textbooks. To avoid reader somnolence (falling asleep), questions, smart and naïve, are asked by a sometimes annoying character called the ***"alert reader"*** who shares them with a twinkle in the eye. *Information concerning the alert reader is set in **fat italics**.*

This book can be used as a source of technical information, it can also be read like a coffee table book, by looking at the pictures, reading the figure captions, and skipping over the technical parts. Where no reference is shown, the source is work – published or unpublished – by Synopsis (see http://www.synopsis.org/index1024/eng/indexeng.html, 1997–2006).

Terminology

The term solar thermal energy (sometimes also labeled direct use of solar energy) denotes:

- The transformation and use of the thermal energy of the sun's incoming rays, by any means of heat transfer (e.g., transmission, reflection, absorption, emission, conduction, convection, phase change, heat storage…)
- The use for any application, be it thermal or electric.

It can be somewhat confusing to find the generation of electricity by thermal techniques (such as thermodynamic cycles in power plants) listed under solar thermal energy, whereas the term solar electricity denotes only PV electricity generation. However, this terminology, as they say, avoids fastidious cross-referencing, and is therefore adopted here.

*Also, the **alert reader** (you will meet her or him quite regularly in the following pages) might notice that isolated parts of text are set in **fat italics**, indicating that the corresponding text is a personal opinion and should be taken, as my father said (and the Romans before him), "cum grano salis," with a grain of salt.*

Introduction: Solar Energy

Sudden events, like the BP Deep Water Horizon disaster and the uncontrollable effects of the mega-catastrophe in Japan, and long-term developments, like the steadily growing awareness of the climate issue and of the finite nature of fossil energy sources, as well as doubts about the place left by the phasing out of nuclear electricity have created a regain of interest in renewable energies, in particular solar thermal energy and PV (photovoltaic) electricity.

In the past, public interest in the energy issue was fueled as much by political, ethical, and environmental arguments as by technical and economic drivers, and renewable energy was often perceived as a field of believers, with little credibility amongst professionals. Since then, new players have entered the arena: clients and, in many countries, legislators voted serious incentives, from tax rebate to plain subsidy, in favor of renewable energies. Although the pendulum might swing back, there is an enormous privately-driven mobilization of capital, know-how, and outright audacity, resulting in a multiplication of new solar products arriving on the shelves and in the catalogues. This mobilization has taken many observers by surprise. Two examples follow:

- As always, when big issues and important technological decisions are at stake, cost ceases to be the dominating yardstick and killer of all but the cheapest solutions: no one has ever seriously tried to compare television to radio in terms of efficiency. When TV became available, everybody who could afford it just bought a TV set.

- In Germany, it was found that private households invest more in their own renewable energy equipment than in utilities, for total investment. This has led to momentary electricity glut situations, where nuclear and coal-fired power plants were not needed, but, *waste for wastes' sake,* had to be kept operating while fossil electricity was "sold" at negative rates.

The battle of the energies is far from over, but (as cynics would say), it is being fought with astonishing fairness, considering what is at stake: the control of the driving force of the economy. However, so far, the democratic process (or the fear of electoral *déroute)* seems to hold.

Hot democratic decision processes need cool information, just as the necessary changes in the energy sector need all of the available brainpower and intellectual honesty to succeed. It is the opinion of the authors that these changes are possible, but that not all of them will come for free, while others might not materialize at all. Some will: the *terms à la mode* are "low-hanging fruit" and "picking the raisins from the cake," which refer to the phasing out of obviously wasteful and unnecessary practices. Few people are going to miss these, or even notice their disappearance (again, some will).

However, once these easy fruit are eaten, we might run out of soft targets, i.e., once the cheapest measures are taken, more difficult targets will have to be attacked, and priorities will have to be set. In fact, and fortunately, this process is well underway. It would not be wise to limit our action to overdue, and highly lucrative, energy efficiency measures, which could encourage consumers to react to price reductions with higher consumption. The term *à la mode* here is "rebound effect." If this effect is real (some experts doubt this), we might come to regret the low-hanging fruit.

This brings us to the question of priority. What should have higher priority:

- Renewable energies, even expensive?
- Cheap energy savings, even without renewable energies?
- Savings plus renewables?

Most people would spontaneously opt for the third alternative, and we suggest that they are right.

Let us be more incisive: what should be the highest priority option? Energy savings or renewable energy? And: what can energy savings do? By themselves, energy savings cannot deliver any energy service, but they can stretch the time axis, and make fossil fuels last longer. Not bad at all, but not enough: we must make fossil fuels last until we have put into place a durable renewable energy system, capable of sustaining itself, BEFORE FOSSIL FUELS RUN OUT. By then, the energy system must be 100% renewable. This is an ambitious target against which we will be measured. The situation can be described as a fuel lantern running on empty in a dark cave. If we fail to find a replacement in time, we will be in the dark, no matter how hard we save: while we can replace energy carriers by other energy carriers, to, say, run an appliance (a lantern or a cellphone), we cannot save 100% AND run the appliance.

> *Finally, in order to succeed, solar – like other renewable energy technologies – needs to fit into the bigger picture in ways that are efficient, economical, and socially positive. This includes the overall energy system as well as specific economic and even cultural frameworks.*
>
> *Chris BUTTERS, May 2011*

1

The Incoming Solar Radiation

On top of the earth's atmosphere, at the average distance between the earth and the sun, the mean energy density of the sun's radiation ("irradiance"), referred to a surface of 1 m^2, normal to the incoming radiation, is 1.367 kW/m^2. This value is called the solar constant, although it is not particularly constant; it changes with the sun's activity ("sun spots," see Figure 1). This change is so slight, in the order of 0.1%, that it needed satellite spectrometric data to find it. A more substantial change, in the order of 3%, is caused by the geometric changes of the reference, the deviation of the earth's orbit from the ideal circular form.

However, the influence of these variations is minimal compared to the 20 to 40% reduction in irradiance during the passage through the earth's atmosphere, due to the mixture of gases called air, suspended matter (as free radicals, aerosols, and aviation contrails), plus the complicated interaction between these elements, shifting concentration, their stability over time, the presence of different greenhouse

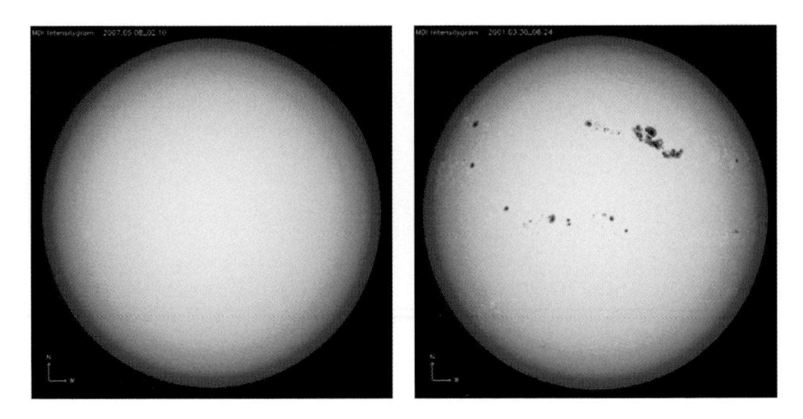

Figure 1 The sun in a calm sun spot period (left) and in an active period (right). (http://lasp.colorado.edu/sorce/newsother/SORCEwebsite_News_Solar_Cycle.pdf).

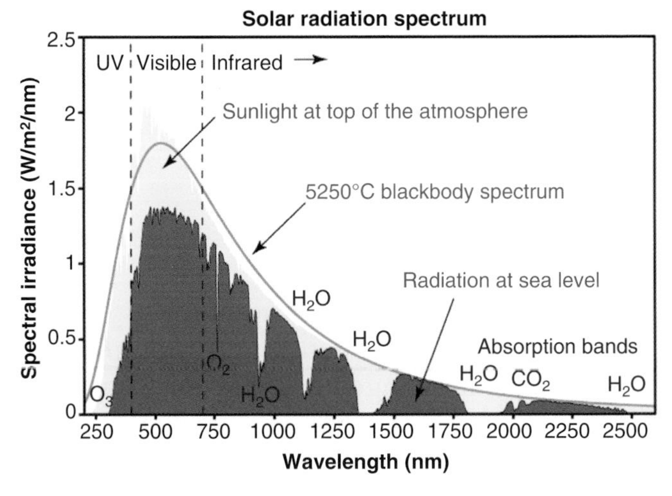

Figure 2 Irradiance spectrum on different levels in the atmosphere (Source Wikipedia).

gases, and solar UV light (photo-smog), all of this whipped by wind and jet stream, sifted through different pressures, temperatures, the effects of human activity and, finally, the eruption of the odd volcano. Figure 2 shows the spectral irradiance (the wavelength distribution of the irradiance). The red spectrum is received at sea level by burned tourists or is available for solar energy applications in two fractions:

- Diffuse radiation having been scattered, but not absorbed on its way. This part is not adapted to concentration, but can be used for low-temperature applications.
- Direct incoming radiation having maintained its original direction on its way through the atmosphere. This part can be used for all applications, concentrating or not.

The difference between the yellow and red spectra is reflected and/or absorbed by the atmosphere. In general, the spectrum marked in red may be available for solar energy applications, provided the sun is shining.

The Availability and Power Density Issue – Fossil vs. Solar Energy

There has been some confusion in the debate on availability and power density, and hence, usefulness, of fossil vs. solar energy. Solar (and particularly solar thermal) energy was often described as a highly diluted, unreliable energy source with limited potential for storage and high-temperature applications, whereas, at present, most experts would agree that:

- Solar energy (and its source, the sun) is a highly reliable, non-depletable energy source... but it is plagued by frequent "meteorological power cuts" (clouds) and limited storage potential; its high and intermediate temperature potential depends on the availability of direct radiation and sufficient concentration ratios; and its potential for low-temperature applications depends on available irradiance and adapted technology, good for tailor-made solutions for specific tasks. It is not particularly diluted: the power of the sun shining on the safety area of a nuclear power plant is in the order of the plant's power, but only during the day... Also, solar energy is prone to heat loss: in order to catch radiant energy, the absorbing element has to be open to incoming radiation, which means also open to potential heat loss, for example, at night.

- The reliability of solar energy can be improved by a number of measures, such as adaptation of supply and demand (e.g., solar air conditioning), storage (heat, dried products), transport (PV super grids), and transformation (synfuels).
- In some respects, fossil energy is the inverse of solar: depletable, but easy to store and uncritical to use at all temperatures. It is practical and versatile, just how much so we realize now that we are running out.

Luckily, many advantages of fossil fuels are also advantages of all fuels (shared by non-fossil fuels); sustainable fuels will remain part of the future energy supply system, as they have been since prehistoric times.

The Need for Tracking

To complicate the solar energy issue further, the sun does not have a constant position in the sky, which is rather good for us (otherwise we would not be here), but bad for solar energy devices. They must be tracked in order to function at optimum efficiency. Also, tracking must be more precise with increasing concentration ratio (and temperature).

The quantitative implications can be found in tables (Solar Energy Pocket Reference, Christopher L. Martin, D. Yogi Goswani, ISES 2005) and websites (www.nasa.org).

To summarize, it can be concluded that the sun's position in the sky is surprisingly different for different places on the globe, and it is standard practice to fool seasoned overseas solar visitors by asking them to indicate due north without a compass.

Of course, this would NEVER happen to our alert reader who, in the meantime, will have realized that the solar vs. fossil issue is getting more complex with every line, and that the voice level is increasing. This is the point where cool information, based on the necessary understanding of details, can help the reader to arrive at his or her own conclusions.

You are welcome.

2

The Basic Solar Energy Heat Transfers

The following first part of this volume presents the technical outlines and functions of heat transfers in solar energy applications.

The angle of description chosen here is to keep track, ad hoc, of the chronological order of heat transfer events facing the incoming light, going through potential or actual transformations, getting:

- Transmitted
- Absorbed
- Used
- And, finally, dissipated.

Put in a more familiar way, we will take a look at the transformations of a light ray, getting caught and used in a solar energy device as heat.

Qualified decisions on the role of solar energy in our future energy system should be based on a shared understanding of the potential and limitations of this technology, which includes the corresponding energy, particularly heat transfers.

Heat Transfer – Experiment and Simulation

Heat transfer can be understood as the metabolism of thermal energy, the exchange of heat between hot and cold. This description can be more or less detailed, in terms of spatial resolution (microscopic, down to the molecular level) to macroscopic; it can treat individual heat transfer modes or

A more traditional description of individual heat transfer modes (conduction, convection, and radiation) can be found in specialized works, such as *Principles of enhanced heat transfer* (Book) Webb, Ralph L, New York: John Wiley & Sons, Inc, 1994 (*ref* 3). As already mentioned above, a typical heat transfer chain in solar thermal energy involves a number of radiative (optical, infrared (IR)) and other heat transfer mechanisms and their interactions. In general, this transfer chain starts with a light ray being ejected by the sun, partly admitted or rejected by control devices, reflected by internal and/or external reflectors; absorbed (transformed into heat) by absorbers or receivers; re-emitted as infrared radiation, reflectors or concentration devices, transmitted by glazing or other transparent components; re-absorbed, cooled (and partially lost) via conduction, convection and other heat transfer modes, used for (hopefully) useful purposes demanding heat or mechanical power; and, finally, dissipated in the environment, i.e., emitted and most likely to end up in the great sea of infrared (IR) radiation bathing the universe at a "cool" temperature of 3K.

compound modes, consisting of several individual, up to global modes. Except for particular cases, this description is empirical (as opposed to derived from first principles); often, it is based on a combination of educated-guess selection of the most decisive transfer mechanisms, simulation techniques, and experimental verification of the simulation parameters and results.

It is important to realize that using solar energy is not an invention of our generation, or even of the technical era. We will bore you by repeating this point several times throughout the book: solar energy is not necessarily linked to a technical device. Traditional grape-dryers or hay-dryers are not much more than a space under the sun and a fork. The principle of some modern applications is barely more complicated: see a schematic diagram of a conventional flat plate solar collector in Figure 3a to get the point.

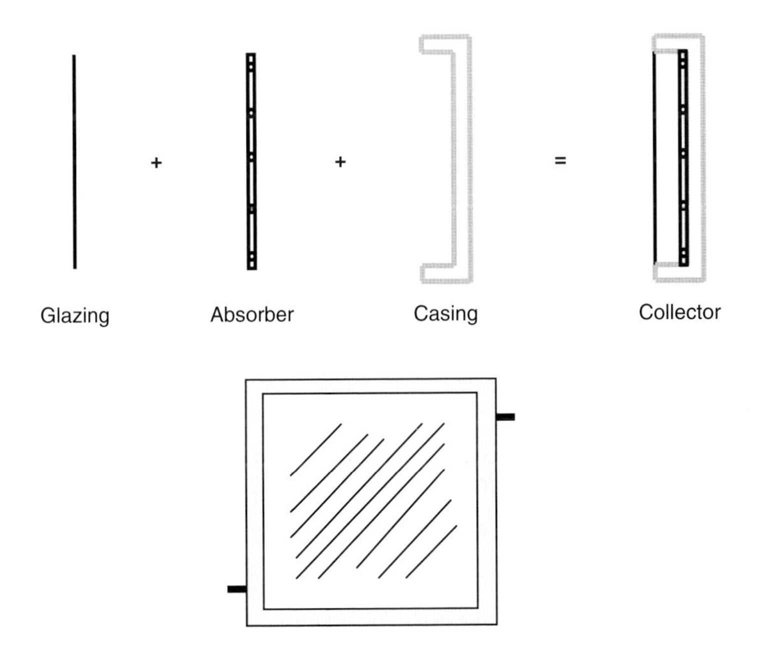

Figure 3a Schematic drawing of the three basic components of a conventional flat plate collector.

This type of collector is used to collect low-temperature solar heat, e.g., for solar hot water heating; it consists of 3 main components:

- Glazing, transparent for the incoming irradiance, absorbing for IR, to keep the heat in,
- Casing, an insulating box frame, with one open side, which receives the glazing (see below), and
- Absorber plate, absorbing for the visible light spectrum, in conductive contact with coolant tubes, to remove the useful heat (e.g., to heat water)

It can be seen that, even with only three main components, the structure of the different heat transfers can be complex. In particular, the incoming radiation has to clear the hurdle of the glazing without being absorbed or reflected (there are two occasions for being reflected, one on each side of the glazing), and it has to avoid being reflected by the absorber surface in order to get absorbed and transformed into heat, and squeeze through the different thermal conduction "bottlenecks" (the analogy with traffic jams is quite adequate) before it can deliver heat across two or more contact surfaces (called "boundary layers," where the main heat transfer resistance is located), across into the stream of transfer fluid, only to repeat the steps in the thermal chain in reverse order, to be able to use the heat on the other side of the heat exchanger.

So much for a "simple" device. The simplicity gets lost some more if you consider the many functions a collector has to fulfill to become a practical, durable and cost-effective device (see Figure 3b).

The "convective" concept has been chosen for this demonstration for three reasons:

Figure 3b Modern flat plate collector array (photo Phoenix Solaire).

- it applies to practically all of the different heat transfer mechanisms (see preceding chapter),
- it has been used for a wide variety of applications, and
- the concept has been published in refereed journals (e.g., M. Grupp, H. Bergler, J.-P. Bertrand, B. Kromer, J. Cieslok, "Convective Flat Plate Collectors and their Applications," in *Solar Energy*, Vol. 55, No.3, pp. 195–207, 1995). In the form of physics, the "convective" setup can be interpreted as a radial counter-flow heat exchanger: the hot air (heated by the absorber) is confined to the well-insulated back of the collector, and enters in contact with the glazing (where most of the heat loss takes place) only after having transmitted the useful heat to the heat exchanger.

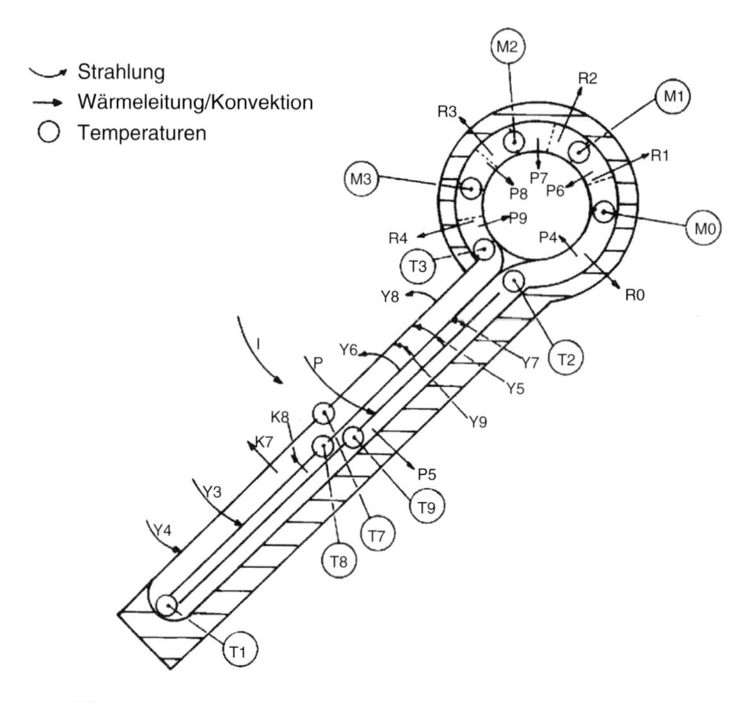

Figure 4a Historical 2-D heat transfer "map" of a "convective" mono-bloc solar water heater, (Cieslok Grupp 3), an original solar water heater (SWH) concept, using a variety of different, liquid and air-based heat transfers. Small circles show "nodes" between experiment and simulation comparison points in this model: curved arrows show absorption or emission, and straight arrows show convection/conduction compound transfers. The big circle (top right) shows a cut through the finned horizontal hot water tank, the collector part (bottom left) a two-way air-type collector: the air in the collector case circulates by natural convection, heating up on its way from T1 to T2 by contact with the absorber surface Y7, heats the tank heat exchanger on its way from T2 to T3, and returns to T1, retrieving heat that would otherwise be lost (Jochen Cieslok, 1984).

It is interesting to insist on the difference in precision between the single efficiency measurement (poor) in histogram and the linear fit curve (excellent).

In practice, all of these different heat transfer modes correspond to as many characteristics to be optimized, a next-to-impossible task for a purely experimental procedure, particularly in the changing environmental conditions of outdoor testing. Results can be obtained by combining simulation and experiment in a more efficient way, compared to purely experimental or simulation approaches.

This is how the simulation part works:

Starting from one of the potential starting points, say, the air temperature T1 in the lowest part of the collector, we calculate:

- the different gains and losses from the absorber and other contact surfaces, using approximate initial values for the corresponding transfer coefficients and mass flow,
- a first approach for the air temperature T2 at the outlet of the absorber,
- and we calculate, to close the cycle, a new value for T1 which replaces the initial value in an "iterative" procedure which can be fine-tuned by comparison with experimental values,
- and finally, the air temperature value of T3 of the air leaving the heat exchanger part.

In this way, if the procedure converges on the same values, independent of the choice of the initial values (which is not always the case), an optimum set of characteristics can be obtained for the description of a wide range of conditions.

The reader who is not yet familiar with the details of the corresponding heat transfers is asked for a little patience. These transfers are discussed in more detail below.

The alert reader might be tempted to ask unpleasant questions about the precision of such a simulation procedure. The answer has been given by an empirical technique, with an exotic name: "Monte Carlo."

Figure 4b presents some examples of histogram results of what is called the "Monte Carlo" technique applied to the Mono-block Solar Water Heater (SWH) shown in Figure 4a: the measured input parameters (in this case, the air temperatures T1, T2, and T3, after convergence of the iterations) are varied following a stochastic procedure to fit the experimental error (scatter). The question is now whether this scatter shows an acceptable signal-to-noise ratio, and, if yes, whether the resulting output distribution shows a reasonable dependence of the experimentally controlled input value distribution. The results in Figure 4b show the variation of the different entry parameters. It can be seen that the experimental values can be reproduced with a precision in the order of 1%, which corresponds to experimental error. Thus, the simulation can be tested for stability and precision.

However, sometimes the practical difficulty of convergence subsists, and requires some patience and feeling – the subjective side of simulation.

Readers who are not yet familiar with collector efficiency curves are requested to keep their patience until the chapter "Overall Heat Transfer."

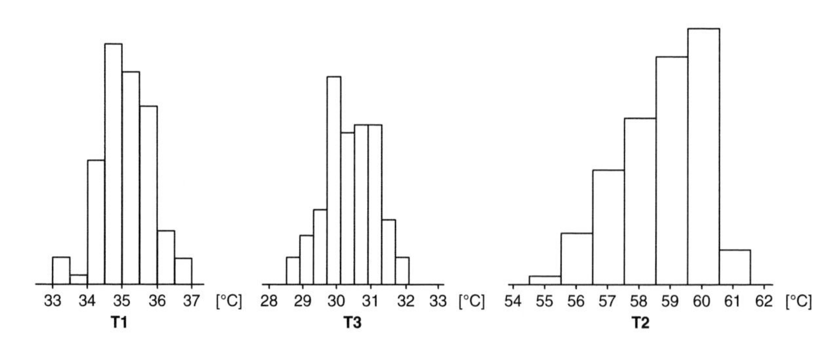

Figure 4b Examples for histogram results of what is called the "Monte Carlo" technique (Jochen Cieslok, 1984).

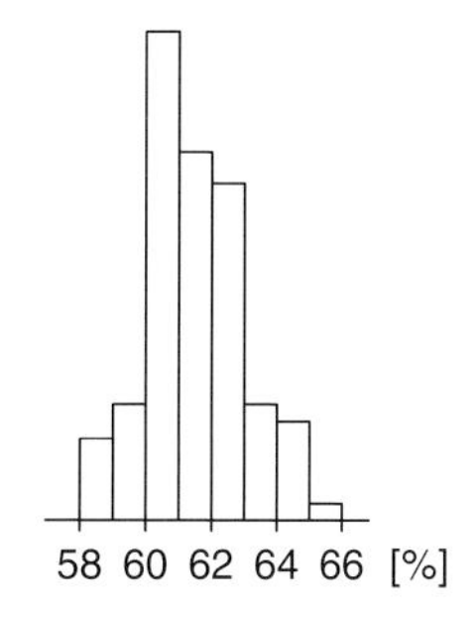

Figure 4c Monte Carlo – effect of input parameter variation on system efficiency. Variation by 1% of the three entry parameters T1, T2, and T3 (in K, see histograms on the preceding page) causes the output efficiency to vary by 3%, which corresponds to experimental error (Jochen Cieslok, 1984). The complete efficiency curve is shown in Figure 4d on the following page.

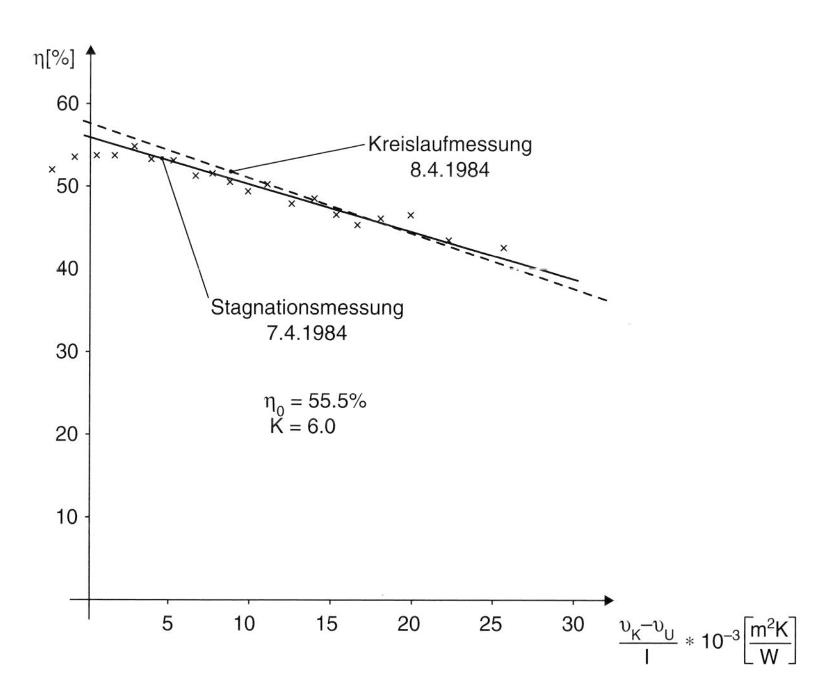

Figure 4d Historical mono-block efficiency curve (Jochen Cieslok, 1984).

The alert reader might wonder whether all this is worth the trouble, just for the evaluation of solar system efficiencies. *Seen in a historical perspective, we are witnessing the takeover of tool over trade. In the past, with traditional experimental equipment (the "tool"), high-level mathematics, such as Bessel functions and differential equations (the "trade"), were necessary to obtain precise results. In the future, probably cluttered with more experimental equipment than today, number crunching might replace our last remaining formal reasoning abilities – until the grid goes off and everybody wanders in the dark wondering what went wrong... Unless opportunity helps to create a countermovement of scientists, like surfers who are used to playing with the forces of nature in an efficient way, hiding their smiles about the helpless "gear kooks" (which the same kooks would translate into "equipment show-off beginners").*

Some more traditional observations on the efficiency curves in Figure 4d:

- Whereas the scatter between individual efficiency values is relatively high, the linear fit parameters between series show excellent precision, even comparing dynamic (figure code x) and static (figure code o) runs, at different outdoor conditions.
- This cannot only be used for the precise determination of efficiency parameters, but also, inversely, for the determination of the other parameters of the linear fit curve (irradiance, thermal mass, mass flow rate, and ambient temperature), or the coherency check between these parameters.

Note that "modern" designs (as the SWH in Figures 4a to d, as opposed to a hayfork) can also be quickly outdated by even more modern designs. The recent availability of sophisticated low-cost components, such as vacuum tube collectors, has changed the market in favor of designs that, arguably, in the higher temperature range:

- are easier to apply (each tube is coupled to the heat transfer circuit in the same way),
- are cheaper to assemble,
- produce solar heat cheaper,
- but need higher up-front investments in R&D (see Figure 5).

Figure 5 A Chinese mono-block SWH with vacuum tube collectors (Photo Onosi).

Solar Energy Heat Transfer Modes

The following part of the book describes the different heat transfers of importance in solar thermal applications, in three different ways:

- Individually, i.e., one by one, as in optical transmission or admission control
- Grouped, i.e., several heat transfers together, as in concentration (reflection and tracking) or convection (conduction and convection)
- Global, i.e., all the significant heat transfers of a heat transfer chain put together in one global heat transfer, as in the global transfer described by the linear collector equation.

Individual Transfers

Admission Control and the Incidence Angle Modifier (IAM)

Let us come back to our incoming light ray. The first hurdle an incoming light ray might encounter on its way to a solar thermal energy system is:

- an admission control device at the entry of the solar thermal system, reducing incoming irradiance in order to avoid overheating or to stabilize input power; this can also be useful for corrections or reference in efficiency measurements in unstable solar conditions.
- Foldable "booster" mirrors which can be closed to protect sensitive glazing material from environmental damage, as well as to protect the internal components from overheating, which, in turn, can lead to thermal damage or evaporation losses by drying out heat transfer circuits.

In many cases, IAM such as shading or total reflection by glazing, as well as other input reduction mechanisms, are defensive, wasteful, and should be avoided if efficiency is the objective. However, IAM can help to stabilize output efficiency measurements in unstable solar input conditions, as in the case of input angular modifiers where an elegant solution consists of measuring irradiance on a reduced model (a custom pyranometer, see Figure 6) with the same angular dependence as the original appliance to be tested. Referred to a normal pyranometer with a flat angular response curve, the IAM pyranometer can filter out the angular dependence of the measured efficiency. This certainly cannot produce high precision absolute test results, but can be used for comparative tests in changing outdoor conditions.

Concentration

One of the most important system characteristics of solar thermal systems is the concentration ratio Rc, a dimensionless variable defined as the intercept area S of the incoming

Figure 6 An IAM-pyranometer prototype: this reduced-size (if not elegant...) model of a SK "deep-focus" concentrator cooker has an optical angular response like the original SK, and a linear thermal response, due to a massive bolt cooling the target to temperatures below the dominance level of IR radiation (see the section on conduction).

radiation on a surface, perpendicular to the incoming radiation, divided by the aperture of the receiver or absorber surface A:

$$Rc = S/A \tag{2.1}$$

Considering that S is the reference surface for the solar gain or, in other words, the surface through which the solar energy enters the system, and A the reference surface for thermal losses, through which energy is lost, a high concentration ratio Rc corresponds to a large energy entry surface, facing a small loss surface, and therefore a high potential operating temperature. However, this temperature has a limit: the surface temperature of the sun (5250°C, see Figure 2).

Rc can vary from less than 1 (for non-concentrating systems such as "high collar" flat plate collectors and pool absorbers) to extremely high ratios.

The Limit of Concentration

No doubt, the local temperatures in an incandescent light bulb can be higher than the above-mentioned 5250°C.

The alert reader might be tempted to ask why a mere light bulb, with indirect means such as electricity, can do what the mighty sun cannot do, directly, by thermal means. If you confront a physicist with this question, he or she will probably answer in a simplified way and refer to the second law of thermodynamics, in about the following way: put two glass bottles in a box, one – in the role of cold extraterrestrial space – filled with cold water, the other – in the role of the sun – with hot water... If you observe the temperatures, you will end up always with lukewarm water, never with scorching hot and freezing cold...Or, smarter, the physicist might observe that the sun can, and does, produce plasma temperatures much higher than the ridiculous 5250°C of a light bulb, but at smaller energy densities than the pretty cool optical solar surface - where the solar power comes from. Sorry bulb...

Tracking

Another characteristic of concentrating systems is the necessity to follow the sun's position in the sky ("tracking"), i.e., changing the alignment of the system at more or less regular intervals. The tracking interval is short for highly concentrating systems; for solar towers, with their large fields of heliostats concentrating the sun's rays onto a relatively small receiver, tracking is a continuous, precise process.

Also, tracking has to be more precise if the receiver is farther away from the heliostats: otherwise, the highly concentrated radiation will be diffused by imperfections in the heliostats and in their support structure. There is ample reason to doubt the story of Archimedes setting fire to the Roman fleet at Syracuse from a safe distance by means of solar reflectors.

Tracking plays a less critical role in smaller systems with lower Rc (see Table 1). This limits the practical size of solar tower plants. Tracking intervals can be longer in the case of linear trough concentrators, some as long as several hours, which allows for manual tracking once or twice a day (as in the case of some parabolic concentrators, concentrating solar cookers, and plane "booster" reflectors for flat plate collectors).

Optical Transmission

Optical transmission is the process through which sun rays overcome partly transparent radiation barriers, such as glazings. It is expressed as the quotient of transmitted divided by incoming radiation energy. Transmission can be:

- Specular or "beam" radiation (the direction of the transmitted light remains unchanged) and/ or
- Diffuse (the radiation barrier re-emits the light equally in all directions).

Table 1 Tracking requirements of different systems – orders of magnitude.

System	Rc	Practical Temperature Limit (°C)	Tracking Axis	Tracking Interval	Unit or Modular
Tower	1000	3000	2	continuous	Unit
Trough	100	500	1 or 2	minutes	Modular
parabolic concentrators	50	300	2	minutes	Unit
non imaging systems	5	150	1	hours	Unit
Boosters	3	150	1	hours	Unit
Vacuum tube collectors/ CPC	2	200	–	–	Modular
Pool heating	0.8	25	–	–	Modular
Flat plate collectors	.9	150	–	–	Modular

In most cases, such as in flat plate collectors, the radiation barrier is a transparent glazing, made of glass, rigid plastics, or transparent film. In low-temperature applications, transmission, together with absorption (see below), has a dominant impact on both efficiency and maximum operating temperature. In receivers under high temperature conditions, typical, e.g., for cracking of hydrocarbons, the solid cracking products in suspension can also act as a radiation barrier. Liquid radiation barriers are found in salt gradient and other solar ponds. Table 2 shows typical transmission and other characteristic values for some solar materials.

Solar transmission materials have to fulfill multiple functions, some of them quite contradictory. Apart from optical transmission (they have to be able to admit a large part of the incoming light), they have to:

- be insulating (have low thermal conductivity),
- withstand high temperature differences and stagnation temperatures, gravity, wind forces, and – in many cases – UV light and chemicals, without degrading over the useful life of the material, and
- be dimensionally stable.

Table 2 Characteristics of selected glazing materials. Light blue denotes generic data.

Material	Optical Transmission Value	Heat Transfer Coefficient (W/m²K)	Maximum Temperature (°C)	Comment
"Green" glass	0.87	5	Up to 500°C	
Low-iron glass	0.91	5	Up to 450°C	Requires tempering
Poly-carbonate single	0.9	5	120°C	Tends to sag
Poly-carbonate double	0.8	3	120°C	Tends to sag
Poly-carbonate multiple	0.72	1.2	120°C	Tends to sag
PE film	0.9	5	100°C	Is UV sensitive
PTFE film	0,95	5	–200 to 260°C	Initial shrinkage

Other requirements are the protection

- of the interior (particularly reflecting and absorbing) surfaces, against normal operating conditions, and
- of the solar energy system against environmental damage, dirt, grease, sand, etc.

Ideally, they should be "self-cleaning" in the local climate they have been designed for, not have a tendency to "pump" in windy conditions, be easy and uncritical to clean, be impact-resistant, easy to install, lightweight, and not create environmental problems – which includes problems after useful life by cluttering the landscape with ugly decomposing material.

Needless to say, there is no clear winner in the competition for the ideal glazing material. But there are some strong contenders. Concerning not only the glazing material itself, but also the corresponding installation systems and tools:

- Glass has been used since the emergence of civilization – for a long time – and can profit from the massive accumulated experience of the building industry.
- PTFE and similar materials, as film, and as rigid single wall and multiple wall systems, with the accumulated experience from agricultural and architectural greenhouses.

Also, prices of the systems differ and can be hard to compare (due to different life-cycle costs), as do esthetics.

Most of these remarks also apply to reflector materials (see Table 3 below).

Reflection

Reflection is the inversion of the direction of (in the solar thermal context, mostly optical) radiation. Reflectivity R is

Table 3 Reflectivity of selected reflectors.

Material	Optical Reflectivity Value	Comment
Silver	0,94	Single reflective surface, on front
Silver, on low-iron glass	0,88	Reflecting surface on back
Anodized aluminum	0,82 to 0,95	Single reflective surface, on front
Polycarbonate, single, aluminized	0,81	On back
Polycarbonate film, aluminized	0,72	On back
Polyester film, aluminized	0,9	Is UV-sensitive
PTFE film aluminized	0,80	Initial shrinkage at first heat up

being defined as the ratio of reflected divided by incoming irradiance. Typical values for reflectivity range from R = .87 to .95 in the visible spectrum.

There are two types of reflectors:

- Reflectors can be single reflective surfaces, such as polished metal. The advantage being higher reflectivity; inconvenience, an unprotected reflective surface.
- Reflectors can also be based on a transparent support, such as glass, polycarbonate, or transparent plastic films, with vacuum deposited metal (silver, aluminum…). The advantage being a protected reflective surface. http://alanod.com/opencms/opencms/Miro/index.html?__locale=de.

Solar reflector materials have to fulfill multiple functions, similar to the corresponding functions of glazing materials (see above), although reflectors are more sophisticated than simple glazing materials. The most critical application for reflectors is heliostats (see discussion in the corresponding section on concentration and tracking). The main reason for this is the higher quality and precision requirements: only specular and precisely reflected light can contribute to power in the receiver on top of the tower. For this application, silver-coated 0.5 mm white glass, mounted on apartment-sized tracking structures, is an excellent choice.

Similar results on lighter, modular structures can be reached with aluminized PTFE foil fixed on circular "drum frame" dish concentrators, with the focal length controlled by slight air depression on the back of the structure.

For smaller, low-cost, low-temperature applications, geometric precision is usually less important than wide-angle reflectance. Anodized (or otherwise corrosion-protected) aluminum, typically 0.7 mm, is widely used, also for reasons of cost and low weight, mechanical stability, and stability against abrasion, fat, steam, and a wide range of chemicals.

For *low-low* cost (but also low-durability) applications, nothing can beat aluminum household foil.

See Table 3 for selected examples.

Conduction

Conduction is the transport of heat in materials labeled "conductors" for highly conductive materials, or "insulators," for weak conductors. The distinction between conductors and insulators is not clearly defined. The conductivities of common solar materials are listed in Table 4. Conduction is, together with radiation/absorption, the only individual heat transfer mechanism concerning solids. If other mechanisms are suppressed (e.g., by the choice of temperatures where radiation is negligible), conduction is highly linear, but less

Table 4 Conductivities of selected materials (Prof. Dr.-Ing Renz: Grundlagen der Wärmeübertragung, 1983).

Material	Conductivity (W/mK)
Silver	410
Copper	385
Aluminum	229
Steel	43
Stainless steel	16
Mercury	8,2
Plate glass	0,8
Water	0,56
Hydrogen	0,18
Water vapor	0,021
Air	0,024

efficient than other mechanisms when it comes to bridging large distances using small heat transfer sections. This makes it interesting for small-scale heating and cooling, and for high-precision applications such as pyranometer (solar irradiance meters, see applications section), but hopeless for distributing heat in central heating systems, compared to convection circuits. However, most heat exchangers rely on conduction across containment walls and the corresponding boundary layers (see below), in fluids on both sides of the containment walls, where sections are large and distances small. ***The alert reader would say that nobody would use power cords to conduct heat.***

Convection

Convection is heat transfer by transport of gaseous or liquid material, the "heat transfer fluid." This fluid is heated on one side and cooled on another, resulting in forced mass flow. If the fluid transfer is driven by thermal expansion causing buoyancy changes, the process is called "natural" convection or thermo-siphoning, as opposed to "forced" convection, in cases where mechanical accelerating devices are used. Closed convection circuits are termed "indirect," as opposed to "direct "or "open," for single pass systems.

Compound or Grouped Heat Transfer (CHT)

Compound heat transfer describes several individual heat transfer modes as one single "compound" transfer mode.

CHT: Convection and Conduction

Convention and conduction are at the basis of one of the most common solar energy components: the heat exchanger, which keeps heat transfer fluids separate, for matter, while transferring heat. The performance of the corresponding heat transfer depends on:

- Flow speed
- Flow type (laminar or turbulent)
- Geometry
- Phase state of the fluid (gas or liquid)
- Phase change (evaporating, condensing)
- Surface affinity, which determines the heat transfer through the boundary layer.

Table 5 shows the case of slow convection heat transfer through a transfer surface, heated on one, and cooled on the other side by fluids. The table confirms that in case of a multiple heat transfer in series, the heat resistance is dominated by the weakest link. Heat transfer improvements should therefore concentrate on the weakest link(s), instead of, e.g., the average heat resistance, or, even worse, on the best heat transfer in the chain. It is interesting to

Table 5 Convective transfer between a primary fluid and a secondary fluid, separated by a transfer surface. The transfer takes place through two boundary layers.

Primary Fluid	Transfer Surface	Secondary Fluid	Total Transfer Coefficient (W/m^2K)
Water	Cast Iron	Air	7.9
Water	Mild Steel	Air	11.3
Water	Copper	Air	13.1
Water	Cast Iron	Water	230–280
Water	Mild Steel	Water	340–400
Water	Copper	Water	340–455
Air	Cast Iron	Air	5.7
Air	Mild Steel	Air	7.9
Steam	Cast Iron	Air	11.3
Steam	Mild Steel	Air	14.2
Steam	Copper	Air	17
Steam	Cast Iron	Water	910
Steam	Mild Steel	Water	1050
Steam	Copper	Water	1160
Steam	Stainless Steel	Water	680

compile literature data on flat plate heat exchangers and to check the influence of fluid and wall material choices on the compound transfer coefficient:

- Air-air gives the weakest results whatever the wall materials
- Air-water and air-steam do only marginally better

- The replacement of air by water on both sides improves heat transfer by more than an order of magnitude
- Water-steam improves water-water heat transfer by another factor of 3.

However, remarks the alert reader, experience shows clearly that it is risky to choose or to discard design options on the sole basis of transfer coefficients. There can be excellent reasons to override quantitative arguments. An example is air, with its low heat exchange coefficient: air is cheap, available almost everywhere on the earth's surface (even where there is no water), not sensitive to leaks or freezing...the air-cooled "Beetle," Porsche (or Corvair) sales record is (still) living proof. And who says that a hot car cannot afford a refitted engine from time to time?

More seriously, a number of air-cooled solar applications have been developed, in particular for desert climates, including solar heaters, water heaters, cookers, and sterilizers (see below).

Furthermore, the term efficiency is relative and implies a reference to an ideal solution, always in danger of being toppled by an even *more* ideal solution, which might make extra heat exchangers obsolete. Finally, the smallest heat exchanger might be the most efficient, but not necessarily the best: if a system already includes a large transfer surface, it makes little sense to add another, certainly more efficient, but unnecessary heat exchanger. Experience shows that air as transfer fluid, in optimized systems such as in Table 5, can arrive at efficiencies comparable to water system efficiencies. See next section.

The boundary layer is the surface where fluid touches the heat exchanger wall, where fluid molecules stick to the wall, and where the main heat transfer resistance is located. Heat transfer across the boundary layer is taking place by conduction.

For a first rough estimate of heat transfer coefficients, the following orders of magnitude can be used, as in the following table.

Table 6 The performance of a conductive/convective heat transfer also depends on pressure and temperature, viscosity, and the flow state of the fluid: turbulent or laminar.

Fluid 1	Fluid 2	Heat Transfer Coefficient (W/m^2K)
Air	Air	5
Water	Air	10
Water	Water	500
Water	Steam	1000

The last point still is a mystery to physics, although attempts were made to understand what is going on when a water tap is opened: a first ordered "laminar" phase sets in and a second phase appears, the flowing water looking like strained muscle: the turbulent phase. In this phase, the flow resistance of the fluid the takes a big step, but nobody completely understands how, when, and why.

Heat Capacity: Phase Change Materials (PCM), Heat Storage and "Thermal Mass"

Phase change, an extremely efficient heat transfer technique, functions by reversible evaporation/condensation (see Figure 7 below), or melting/freezing cycles. Evaporation/condensation cycles taking place in evacuated, "one-dimensional" (as opposed to closed-circuit) tube geometries, in which the steam transport takes place in the same tube as the condensate return flow are called heat-pipes; they transfer heat quasi-isothermally, i.e., without changes in sensible temperature. This is interesting for solar thermal applications, which tend to operate best at lowest

Figure 7 Eutectic salt: A mixture of two melting salts with different melting points. Used for high-efficiency cold storage (photo: Chris Butters).

temperatures. In these cases, slightly higher temperatures mean substantially higher losses.

Melting/freezing cycles operate best on demand (particularly in cooling applications in the presence of a solid close to melting temperature). The solid phase implies relatively high heat transfer resistance at constant operating temperature.

More versatile, but not as efficient, are controlled storage tanks, insulated and operating with a liquid thermal fluid, storing sensible heat. A tank with volume V, storing a fluid with a specific density D, and specific heat H, stores thermal energy Q when brought from temperature T1 to T2:

$$Q = V * D * H * (T2\text{-}T1) \tag{2.2}$$

Introducing the term "thermal mass" Mth:

$$Mth = V * D * H \tag{2.3}$$

Figure 8 Outdoor heat storage tank: Lots of insulation for lots of storage.

Absorption and Emission

Absorption is the main destination in the short life of our "successful" light rays: after having cleared all the hurdles, they can dive into the absorber, and, if they are "lucky," one last time, they will get absorbed.

The probability of getting absorbed, called the absorption coefficient or optical absorptivity (a), is highest in the optical part of the spectrum (see Figure 2). After having deposited its heat in the absorber, it heats a fluid, such as water, to a temperature necessary to take a shower. ***Seen from the perspective of the light ray, the alert reader will be surprised about:***

- ***The number of light rays it takes to heat a small quantity of water***
- ***The enormous quantity of light rays that get lost before one light ray succeeds in getting through to the absorber***

- ***How cold and empty it is in the universe where the little rays get emitted to after being "dissipated" (i.e., emitted at low temperature).***

Until the ray finds a celestial body in its path: dust, meteorite, moon, planet, star, the occasional black hole…

Table 5 presents some typical absorption ((a) for the total spectrum) and emission values ((e) for the IR part) for commercially available absorbers.

There are two types of absorbers:

- Non selective where: a = e
- Selective where a >> e.

Selective usually outperform non-selective absorbers in the high part of the operating temperatures, where IR-losses are the dominant loss mechanism. As for absorbing materials, there are often selection criteria other than performance, such as mechanical, thermal, and chemical resistance, in cases where the absorber surface is exposed to a hostile environment. This applies foremost to selective absorbers with cavernous surfaces, i.e., where the effect of selectivity is obtained by microscopic caverns acting as traps for optical radiation with small wavelength ("fitting"

Table 7 Range (bracket) of emission and absorption coefficients of 16 commercially available absorbers (Solar Energy Pocket Reference, Christopher L. Martin, D. Yogi Goswani, ISES 2005).

Selective Absorbers	**Bracket**
Absorption (a)	0,92–0,97
Emission (e)	0,03–0,18
Non-selective Absorbers	**Bracket**
Absorption (a)	0,85–0,95
Emission (e)	0,85–0,95

into the caverns, whereas the longer wavelength IR radiation "sees" only a smooth reflective surface). In practice, it is only after the corresponding durability problems are solved that performance criteria should be addressed.

Convection in Collector Circuits

Natural Convection

Natural convection is based on buoyancy changes due to temperature differences in the fluid: the heated fluid becomes lighter and floats upward. Typical natural convection velocities in small solar energy systems are in the order of 0.2 to 1 m/s. The advantage of natural convection circuits: no need for auxiliary energy for pump and ideal passive control strategy; disadvantages: bigger tubes required, the position of the tank must be above the position of the collector, and no high point in the circuit is allowed.

Forced Convection

Forced convection is the heat transfer by heat transport of fluids, where the fluid is accelerated by pumping devices.

- Advantage: the circuit components can be installed in any position, which allows for a "one-size-fits-all" installation and kit. Higher transfer coefficients.
- Drawbacks: needs pumps, controls, auxiliary energy, creates pump noises, potential risk of night cooling by inverse convection.

Overall Heat Transfer

Collector Equation

In principle, a solar flat plate collector is a simple device: a cooled absorber, glazed on one side, insulated on the other (cf: Figure 3a). You can put it on a rooftop for 20 to 40 years, if you do it right. In the next section, some of the corresponding requirements are discussed. In one respect, the solar collector remains simple: it is expected to produce heat.

Most modern collectors do this, with minor differences in price per square meter, in efficiency, and thus in lifetime energy cost. To determine this prospective cost, different methodologies have been defined. The basis of these methodologies is presented in the following sections.

Output power OP and efficiency Eff of a collector are functions of the following parameters, which are determined or monitored experimentally by the type of sensors shown in brackets:

- Collector inlet temperature T1 and collector outlet temperature T2 (platinum or thermocouple sensors).
- Fluid flow rate: direct (by impellor sensors) or indirect (by temperature difference and thermal-power calibration).

Examples for other parameters to be recorded are:

- Collector Surface.
- Incident Angle Modifier (IAM).
- Ambient temperature (Ta) (by carefully shielded and vented sensors).
- Direct and diffuse irradiance (Irr) (by pyranometer).
- Other ambient parameters: wind, collector elevation, tracking frequency by data-logger, and careful consignment of all data into the test protocol, which can make or break a successful measurement.

The evaluation of these parameters takes place with the help of different versions of the "collector equation" describing the dependence of collector efficiency from collector temperature, ambient temperature, and irradiance. The collector equation is based on total heat transfer coefficients, i.e., for the total Compound Heat Transfer (CHT) of the collector. This has the advantage that the test treats the collector as a black box, without any respect for particular internal function modes. Details on the test procedures for the different parameters can be found in the next section.

Linear Collector Equation

The linear collector equation is not only the simplest way to evaluate collector performance data, but also to calculate parameters on the basis of available performance data (see p. 52).

Two simplifying assumptions are made (for an excellent presentation, see David Faimann in http://physweb.bgu.ac.il/COURSES/EnvPhys/SolPhysLect5(2003).pdf).

1. All heat transfer coefficients, i.e., transmission through the glazing, absorption of irradiance, IR emission, conduction through the boundary layer into the transfer fluid, convection and conduction through the glazing and through the insulation are compounded into one total or global heat transfer (THT);
2. The total heat transfer coefficient is assumed to be linear, i.e., only the linear part (where IR-emission and evaporation are negligible) is included in the evaluation.

The Output Power OP of a flat plate collector can be written:

$$Op = m\ dot * cp * (T2 - T1),$$

with m dot the transfer fluid mass flow, cp the specific heat, T2 collector outlet temperature, and T1 collector inlet temperature. In linear approximation:

$$Op(T') = Op(0) - a1 * T'/Irr$$

With T′ the average absorber-ambient temperature difference and Irr the solar irradiance

$$T' = 0.5 * (T1 + T2) - Ta$$

For the efficiency Eff (T′) of collectors in thermal equilibrium = Op(T′)/Irr, this yields

$$Eff(T') = Eff(0) - (a1 * T')/Irr \tag{2.3}$$

Equation Eq 3 is called the linear collector equation, with Eff (0) the optical collector efficiency at temperature difference T′ = 0 K, and a1 being the linear total loss coefficient in W/m^2K. These two parameters are normally sufficient for the precise ad-hoc characterization of solar energy thermal systems in their linear range.

Since this equation is a cornerstone of solar thermal energy, and will be used further in the framework of this book, in particular for a non-intrusive measurement of pot content being cooked in solar cookers – which allows a precise answer to the eternal question whether solar energy appliances are really used (*see section on Solar Cooker Use Meter, page 93*) – it is worth to take a closer look at it.

It consists of a constant term, corresponding to the solar heat gain which is assumed to be constant – whereas the heat loss is linear in T′, leading to ever-increasing losses, which allows the determination of Ta.

For collectors in thermal equilibrium, the constant input term can be written as:

$$Eff(0)=Fr * TrOpt * AbsOpt \tag{2.4}$$

With Fr the heat removal factor (see below, usually close to, but smaller than 1), TrOpt transmission of the glazing and AbsOpt the absorption of the absorber (see preceding sections), both factors, in modern materials, being close to 90%, this yields, as a rule of thumb:

	Eff(0) = 0.8
The input power Pin is	Pin = S * (collector surface) * Eff(0) * Irr
The loss is	Ploss = K1 * S * (Tcoll-Ta)
Output power Pout	Pout = Pin–Ploss
Eff: the efficiency	Eff(T′) = Pout/Pin

As for the linear loss factor K1, the approximate value of K1=of 4 W/m^2K with low iron glazing, selective absorber, water as primary fluid, and air as secondary fluid can be used as a rule of thumb. This means that, at ambient temperature of 20°C, 80% of the incoming solar energy is transformed to useful heat.

A low efficiency collector (K1 = 5 W/m^2K) still reaches boiling temperature at an ambient temperature of more than of 20°C and an irradiance of more than 500 W/m^2 in linear approximation:

$$\mathrm{Eff} * T'(0) = \mathrm{Fr} * \mathrm{TrOpt} * \mathrm{AbsOpt}$$

and

$$\mathrm{Eff}(T') = \mathrm{Eff}\,(0) - (a1 * T')/\mathrm{Irr}.$$

Second Order Collector Equation

In the high-temperature part of the efficiency curve, most collectors show an efficiency drop due to IR emission and other non-linear phenomena. To obtain better agreement between theory and experiment, a non-linear term has been

Table 8 Typical values of transmission, absorption, heat removal, and linear loss coefficients of different collector types.

Type	t Glazng 1	t Glazng 2	Abs	Fluid 1	Fluid 2	Fr	B (Tr * abs)	Fr*	K
unglazed PE pool absorber	1	1	0,6	Water	–	0,9	0,60	0,54	15
traditional SWH collector	0,87	1	0,9	Water	–	0,96	0,78	0,75	5
traditional double glazing	0,87	0,87	0,9	Water	–	0,96	0,68	0,65	3,5
low iron single glazing	0,91	1	0,9	Water	–	0,96	0,82	0,79	5
low iron double glazing	0,91	0,91	0,9	Water	–	0,96	0,75	0,72	3,5
selective single glazing	0,91	1	0,93	Water	–	0,96	0,85	0,81	3
selective high tec glazing	0,95	1	0,93	Water	–	0,96	0,88	0,85	0,4
high tec double glazing	0,95	0,95	0,95	Water	–	0,96	0,86	0,82	
air unglazed	0,95	1	0,6	Air	95	0,96	0,57	0,55	
air glazed	0,95	1	0,8	Air	–	0,96	0,76	0,73	
air glazed "convective" coll	0,95	0,95	0,93	Air	water	0,94	0,84	0,79	5
vacuum flat plate collector	0,95	0,91	0,8	ht ppe	water	0,96	0,69	0,66	1,3
vacuum tube collector	0,95	0,91	0,8	ht ipe	water	0,96	0,69	0,66	1,5

added in one of the approaches in the fit curve: a square term (which is a function of T'^2) with fit parameters K1 and K2. Although a simple second order dependence cannot reflect the varied heat loss physics it is meant to describe (IR emission and flow resistance in small tubes are forth order heat transfers, whereas convection shows even more exotic qualitative jumps), an ad-hoc second order version of the collector equation has been adopted in European performance test norms. A square term has been added in the fit curve:

$$Eff(T') = Eff(0) - (K1 * T'/Irr) - \mathbf{(K2 * T'^2/Irr)}$$

This term (set **bold** in the equation above) is not easy to visualize. Moreover, the temperature range to which this part of the collector equation applies is close to stagnation temperature, a most critical range for flat plate collectors that designers try to avoid. In cases where this is not possible, explicit simulation of heat transfers in the collector can be used for optimization, as in Figure 4a.

3

Solar Thermal Energy Product Requirements

This chapter goes into some more detailed examples concerning requirements of solar thermal energy for the hardware. These requirements are easily underestimated: the absolute values of temperature and pressure in a SWH are no challenge for the respective materials, but they are recurring for many years, diurnally cyclic, and taking place in a hostile chemical and UV environment, in thermal shock conditions, freezing to stagnation, and subject to hail, abrasive sandstorms, and more or less violent cleaning. One of the authors adapted risk assessment methods developed for nuclear reactors to the risk of failure of solar collectors: he found two important failure paths, one of them (thermal shock by partial obturation of glazing at stagnation temperature) repeatedly occurred, but the manufacturer did not want to carry the cost for tempering. The second failure path had not materialized during the useful life of the collector.

In the 1980s, solar thermal water heating equipment was of very poor quality: intermetallic corrosion, started by condensation, casing leaks and outgassing foam, led to major tank and absorber leaks, after very short service life times. In these pioneer days, quality was simply not an issue, with some exceptions.

Some 10 years ago, it became an issue. In a study on the SWH market in four European countries commissioned by the EC-TREN, it was found that solar professionals in France were very critical about the quality of the products they sold and installed at low cost, compared to German products that had already switched to higher quality at higher cost. This divided the storage tank failure rate by 10, in a comparison between the two countries. While there is no compelling reason to ascribe a cause-effect relation, the fact that tanks did not leak rusty water anymore must have improved the public perception of solar water heaters. Perhaps the situation was more of a syndrome: at one side, improved quality enhanced interest on the technology, which in turn motivated market players to further improve quality of products on the market, which, finally, did not harm the product image, and so forth.

In extreme cases involving large systems, the quest for higher performance might reduce durability, to the point where the use of lower-performance material looks more attractive than the use of higher-performance material. The example of the flat plate collector SWH shows that the optimum solution is not necessarily the highest performance (of collectors, tanks, and of pipe insulation), which leads to higher stagnation temperatures, and, in turn, to prohibitive pressures in the system, particularly during these long summer holidays when the sun is there, but nobody is there to use hot water (or to service the system), and where vegetation gets steamed by glycolized water instead of getting watered.

Clearly, venting off steam is a potential solution to avoid serious fluid loss and the corresponding damage. There are ways to limit consequences. There is an excellent book (Felix A. Peuser *et al. Solar Thermal Systems*, 2002) on the subject, containing the details for the selection and sizing of the components. The main points are, as indicated, low volume, "meander"- or "harp"-type absorbers: these geometries avoid the ejection of liquid water and the resulting shorter service intervals for refilling of fluid with its cost and the impressive discharge and potential accident risk of superheated water:

- Chose absorber and bleeding valve positions ensuring to vent steam (as opposed to superheated water).
- If required, allow collectors to run dry, even if this exposes the system to dry stagnation temperatures of above 200°C.
- Chose high-quality buffer vessels of sufficient volume.
- And, most importantly, organize a reliable and qualified service, including remote sensing and regular visits, and make sure the personnel is not on summer holiday when they are needed in a hurry.

This service will also be convenient for the implementation of the local anti-legionella procedures.

E. legionella is a bacterium which can infest large (volume > 400 liters) hot water tanks when these tanks remain at moderate temperatures (<60°C) for more than a day. Legionella can cause life-threatening pulmonary disease. Smaller tanks and higher temperatures eliminate the risk.

Figure 9 A thick layer of dust can hide qualities of a solar collector.

Figure 10 A vacuum tube array integrated in a traditional roof (Photo Phoenix solaire).

4

Selected Solar Thermal Applications

Solar Water Heaters (SWH)

Solar Water Heaters (SWH) are by far the most used application of solar thermal energy. They exist in a wide variety of system sizes, from simple black plastic bags as camping showers to district installations. The same variety can be found in important differences in system characteristics, such as:

- Thermal performance, particularly under unfavorable weather conditions, which impacts the "solar fraction" (the percentage of hot water needs covered by the SWH).
- Temperature stratification in the tank of the SWH, in priority the upper regions of the storage tank: this can be a desired effect, since a small amount of hot water is available after a short exposure time. Stratification buildup is

caused by heatup convection, while destratification is due to convection, mixing with incoming water and to conduction in the tank walls, as could be shown by simulation and experiment; to a certain extent, stratification can be controlled by inclined or vertical cylindrical tanks, as opposed to horizontal tanks.

- Wasted water, due to cold water in the hot water feeder pipe, forcing the user to run off this water before the arrival of hot water: this is a problem caused by the SWH placed at a distance from the hot water withdrawal points, e.g., on the roof of apartment buildings. There is an elegant solution to this problem: hook up a closed circuit.

The alert reader might observe that this causes important energy losses by the pump. A classic dilemma: lose there, win here. As long as you keep track and steer.

The most important system types are, starting with lowest costs:

- **Integrated passive solar water heaters** (IPSWH), also called batch water heaters or breadboxes (the tank and the collector are the same element), from DIY wooden boxes to super-insulated high tech models: see Figure 11.

These models were flagships of California's ecological architectural antithesis to the mainstream lifestyle of the 1960s and 1970s – *remember color photos of drop city et al. Cost: truck fuel for many trips to the junkyard, time well spent for collective design and building, first encounters with obstinate reality, such as freezing and corrosion damage, re-discovery of temperature stratification in horizontal tanks. Voice from the shower ("this is HOT – now wait – this is lukewarm getting colder...")*

Figure 11 "Post-industrial" DIY passive solar water heater with 2 * 150 liters capacity, insulated booster reflector and counterweight-assisted elevation booster tracking. For reasons of stratification stability, the two tanks should be installed in parallel.

Figure 12 Large integrated solar water heater in Greece.

- **Monoblock** solar water heaters, with absorber and tank being separate components in a common casing.

Figure 13 Flat-plate monoblock SWH on a Greek flat roof.

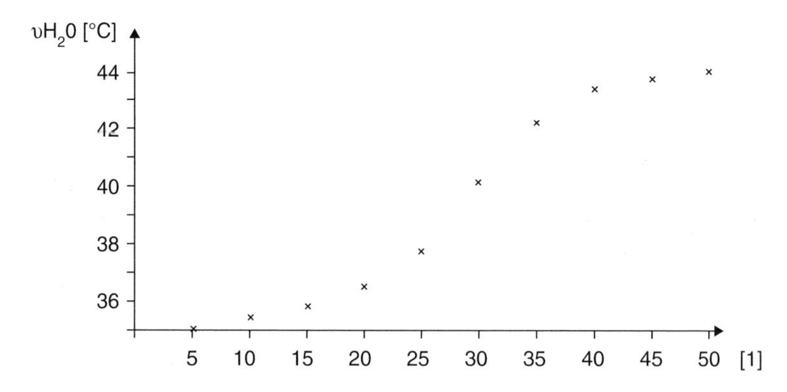

Figure 14 The temperature stratification in the tank of a monoblock SWH of 50 liters (horizontal tank) is determined in the following way: the water is withdrawn slowly through the inlet on the bottom of the tank (cold water first and hot water last). The x-axis shows the number of liters of water withdrawn, in units from 5 to 50 liters. The y-axis indicates the equivalent temperature. For normal withdrawal (hot first), during the first 25 liters, the temperature stays above comfortable "shower level," but changes significantly during withdrawal.

Figure 15 "Split gravity" solar water heater in Eritrea.

- **"Split gravity"** (the tank and the collector are separate elements, placed in different casings). The example in Figure 15 shows a unit with $8m^2$ collector surface and 500 liters tank volume, "open" atmospheric pressure heat exchanger, and after sales-service limited to occasional refilling of circuit. Tank insulation cover by tarpaulin, Tichelmann[1] series-parallels circuit. Advantages: in general, split gravity SWH are arguably the most interesting SWH design by far, with ideal control strategy, stable stratification, and no auxiliary energy needed. But, on the other hand, these systems need to be planned individually, in order to obtain an efficient gravity flow. Also, they cannot be installed

---

[1] Tichelmann would have been forgotten, except for his funny name and for the simple intelligence of balancing flow resistance by balancing series and parallel parts in hydraulic circuits.

in every situation – which needs expertise at an early stage in planning – where cash for planning is often not yet contracted.

- **For "split pump" systems** advantages are mainly the polyvalence (can be placed anywhere, using standard material) and standard budgets. Disadvantages: need of pump, control, and auxiliary energy.

"Split gravity" and "split pump": two different concepts, similar looks, but a huge difference in market success.

And the winner on the market is... mono-block in the low-cost category, and pumped split in the upmarket categories.

The alert reader says: "The market has spoken: matter over mind..."

Solar Space Heating

There is a traditional division in solar energy between "passive" houses (they just sit there, need no pumps, no controls, no auxiliary energy, no tubes, no collectors, and no tanks, although most people would draw the line further up in complexity). The complementary term is "active" (for houses featuring mechanical parts, pumps, controls, etc.). The historical definitions between active and passive systems are presented below.

> Natural energy flows in the environment have always been used to heat and cool us as well as to perform many other tasks such as crop drying, baking.... This includes useful energy from sun, wind, water, and earth. Whereas passive heating is mainly linked to solar radiation, passive cooling often exploits airflows – as well as other ambient conditions in water or the ground. Good passive design is the art of working with all the elements of nature.
>
> Chris Butters, June 2011

Direct Gain

The simplest concept, used by plants and animals, is as old as life itself: it is called direct gain, which includes openings towards the sun such as troglodyte housing and, the ancestor of solar heating, a pad at the opening of a cave. Direct gain is still a main contender today; we shall not see the window replaced any time soon. And the window is still under intense development.

The interest of direct solar gain systems is largely dependent on what glazing technologies are available.

Windows and Glazings in Solar Space Heating

Early windows had single glazing, and hence, often let more heat out at night than they gained during the day, and on cloudy days as well. Various systems with multiple glazings, including special types of glass and coatings, have greatly improved the heat balance of windows, but the cost is correspondingly high. There are windows specially designed to reduce solar glare and to reflect heat in hot climates to avoid overheating, which can be a dominating energy loss path in air-conditioned latitudes. On the other hand, these windows also let in less daylight, typically by 10 to 20%. The corresponding loss path due to higher need for artificial lighting has lost some of its cost-saving potential since the introduction of energy efficient lighting, such as CFL and LEDs.

Multi-layer glazings with argon filling or vacuums are efficient solutions, but have some drawbacks: their cost, and also their vulnerability to puncture or damage, which reduces their useful life; they also cannot be repaired, as old windows could be.

As Chris Butters puts it, despite technological improvements, there is still the basic issue that day and night are different, and that windows lose heat at night; an obvious solution is to insulate windows at night. Although interior curtains help to some extent, to be efficient, the insulation should be external. But this either requires users to go around their building every evening and morning – ***the alert reader is thinking: no more sleeping in late on weekends?*** – or else automated external shutter systems. The latter, however, tend to be expensive, and can become iced up in cold climates.

External roll-down shutters are quite common in some European countries, and are known as "Rollladen" in German. They double as night insulation and summer shading. However, their usefulness is less today, given the latest energy-efficient windows.

There have been some "good ideas," such as fitting windows with a vacuum cleaner motor, blowing Styrofoam (polystyrene) beads into the gap between glazings at night, and sucking them back out the next morning. In this way, the glazed areas could in principle become fully insulated at night. ***The alert reader thinks: "Beware of flying glass – it is a good idea to wear safety glasses when testing this setup."***

Another example from Canada was an insulated panel sliding up in between the glazings. Unfortunately, the area of wall where it was stored in daytime thus became correspondingly less well-insulated at night.

Chris Butters again:

"It is curious that we should be eternally seeking one single technical solution to cover opposite conditions, while a more intelligent approach is to respect nature and treat summer and winter as different; to design *with nature*. One solution is to insulate the windows in winter, as has been the tradition in chilly Scandinavia for decades, and still is quite common. This is done by adding an extra set of casements in winter, and removing them in summer. The operation takes only minutes as well as some solar common sense."

In the future, windows could (but probably should not) be "improved" to be replaced by other, different technologies, such as light ducts, fluorescents liquids, and light-sensitive glass.

Again Chris Butters:

"The glass industry and architects are still looking for ideal solutions, where buildings with acres of glass can be made to perform sensibly in energy terms. This quest may or may not succeed. Certainly, advances are being made, and modern windows lose less heat than old-fashioned single-pane windows. One important factor is that the frames are now insulated, too. But this means that windows with small panes and glazing bars, which increase the heat losses, cannot be used. However, at present, even with the best available windows, it is still the case that the window parts of the building generally lose four times more heat, and also cost about four times more, than a normal well-insulated wall per square meter. Then again, there are the different gains, direct and indirect, mostly (if heating is the main issue) on the sunny side of the building, but also elsewhere (if cooling is the main issue). There are good reasons to use glass, within reason. "

The issue of glazing bars is leading to hot discussions with cultural heritage authorities, who wish to keep the looks of the old windows. Many people also prefer windows with small panes. It has been shown that the energy performance of old windows can be greatly improved by adding new interior layers, which is not quite up to the most efficient standard, but is pretty good. This is where one has to consider the *overall* value, including aesthetic and historical values, and not only our specialist issue of the energy balance.

Table 9 Typical U-values observed in different glazing concepts.

Window Type	U-values (in $W/m^2 * K$)
Old single pane glass	*5,0*
Old sealed frame double panes	*2,0*
Various low energy windows	*1,0*
Best available "passive" standard	*0,6 (Including frame – and at 25–40% higher cost)*

Since the heat balance is more advantageous in the case of sun-facing windows, this is where passive solar heating has an important role to play, provided that one adapts the application to the right climate and utilization. In the far north (or south), the net gain will be considerably less than at more sunny latitudes.

Active and Passive Solar Energy

The terms active and passive are not defined by precise qualitative or quantitative criteria. In a first approach "passive" was solar: it could be anything from wearing a black shirt to hanging out clothes to dry; passive cooling could include sitting in the shade. The terms active and passive were applied in the field of energy-efficient architecture, starting in North America, from the 1970s. ***In these early passive houses (as the alert reader might observe), the building envelope itself did the job of collection, dubious passivity, as well, since walls and south-facing glass are also active technology.*** The general difference is one of intention: the passive solar building is designed like a solar collector to trap and then store solar radiation, as far as possible without any moving ("active") parts, with the active approach being the application of technical components. However, in order to function efficiently, the owners of passive houses often had to close insulating shutters at night and open vents in summer, whereas the active solutions were in principle automatized.

The alert reader might observe that passive houses needed active inhabitants whereas you could put passive people into active houses. The current trendy idea of "smart" houses is similarly based on the assumption of passive inhabitants.

In modern solar houses, the distinction of active vs. passive has lost much of its analytical value: while a passive house in the past relied only on passive elements for its heating, practically all passive houses today feature active as well as passive elements: a passive house in today's terminology is a solar house featuring direct gain elements; very few are purely active or passive houses.

Also, some confusion has been created with a new generation of so-called "passive houses;" this is fast on the way to becoming synonymous for the energy standards for new buildings in many countries. Originally developed in the 1990s in Germany, the term "passive house" refers to a very low-energy building with a specific heating energy demand of less than 15 kWh/sq.m.year. The design is based on four main principles: extremely efficient windows; extra thick insulation; passive solar heating from south-facing windows; and controlled ventilation with heat recovery. In these buildings, the free heat supplied by the users' bodies plus heat from lights and other appliances is normally enough to heat the buildings with no added heating system. Hence, the term "passive."

Figure 16 Cut of a direct gain passive house; the low winter sun gets admitted to the inside of the house where it heats the massive floor; the high summer sun is blocked (drawing Guy Renaux).

However, as the alert reader might point out, these buildings are in fact entirely reliant on a very "active" ventilation system, which is in fact a mechanical heater.

In addition, the passive energy standards as defined to date do not include extremely important factors, such as the embodied energy of the building. But that is another story.

As we have seen, the term passive is therefore anything but rigorous. It is however still useful in describing the intention to use in preference natural energy flows, in particular solar radiation and ambient airflow, preferably in simple ways, and in general using already-existing components while avoiding to add dedicated technical machinery.

Passive Solar Heating and Overheating

"Passive solar heating" relies predominantly on direct gain, i.e., on glazing. However, there is always too much of a good thing: glazed rooms heated in this way will have a tendency to overheat within an hour or two. With an "active" solar collector, the excess heat can be transported away to some storage tank (hence needing the "active" input of a pump) unless natural convection (or gravity flow) can be used. In the case of the "passive" building, it is the walls, floor, etc. storing the excess heat that return it to heat the rooms on the following night: if this function is taken over by heating or cooling circuits, it is called activation of building components, which hints at the presence of "active" components. Naturally, these simple principles are often combined in various hybrid solutions, where part of the heat is used directly, and another part is transported away for later use.

Façade Collector Systems

The building façade was identified quickly as ideal location for space heating collectors: the low winter sun, amplified by snow reflecting from the ground, has first convinced the happy few living in clear winter climates; the others had to wait for the advent of vacuum tube collectors.

The Trombe Wall

One of the true genius solar pioneers, on equal footing with Harry Tabor, who, almost single-handedly, invented modern solar energy technology, is Felix Trombe, the inventor of the Trombe Wall. The southside wall of the building serves as glazed collector and storage element, coupled to the room by controlled convective air-flow passages. These passages are acting, as one could say, as "passively activated components," a long time before the term "activation of components" was coined.

Other Parameters

Temperature and light are far from being the only comfort-related parameters of a building, the alert reader says: there is also humidity: if it is too low or too high, you should seek for serious advice on condensation and related phenomena before moving in with your uninsured Stradivarius.

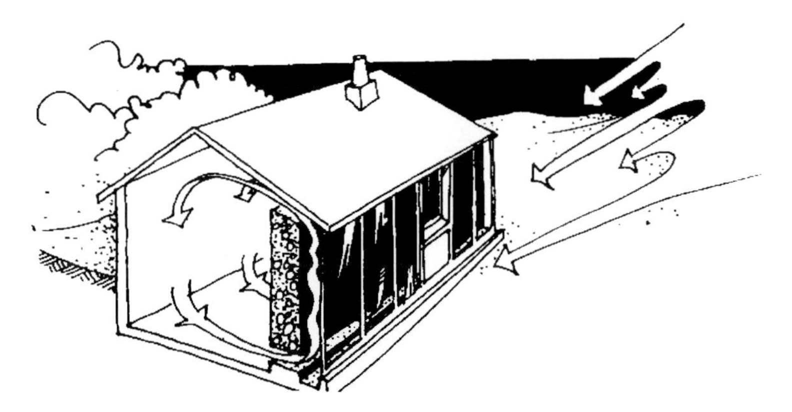

Figure 17 Trombe wall: from right to left: glazing with air intake, light absorbing heat storage wall, gravity flow air passage to the inside of the house (drawing Guy Renaux).

Figure 18 Trombe wall of the Kelbaugh house in New Jersey (Photo Chris Butters).

Other Façade Systems

Another interesting development is an absorber element with the air intake taking place through small holes in the Al absorber sheet: a radial counter flow heat exchanger in its utmost simplicity. However, the applications of this principle are mostly limited to dust-free, high airflow, low-temperature heating requirements, such as special purpose industrial hangars.

- **Thermowall**, a Franco-Swiss development, is an application of the "convective" 2-way air-type (see Figure 20 below) collector, avoiding conventional flat plate collector drawbacks, such as leaks, corrosion, and limitation in module unit size, with comparable efficiency. A multifunction application to high population density situations is being presented at the end of this book.

- **Transparent Insulation Materials** (TIM) were used for passive heating, invisibly integrated into south-facing wall elements of the ISES building in Freiburg, Germany.

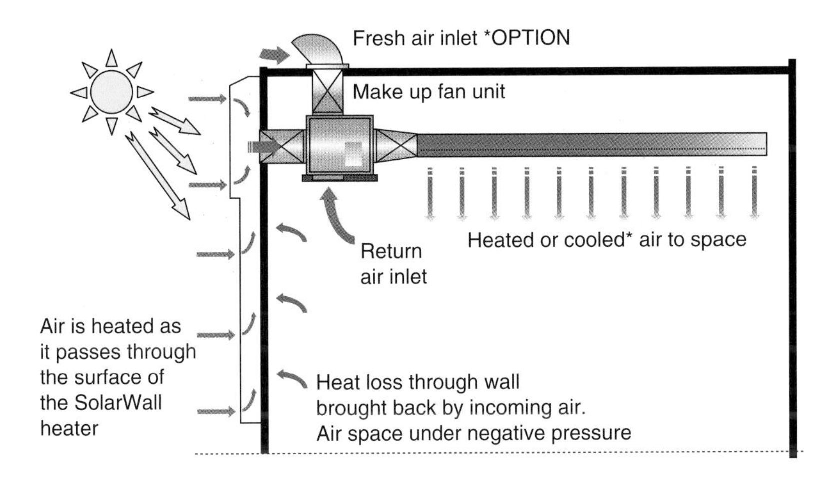

Figure 19 Unglazed solar façade: The incoming air is heated in contact with the SolarWall surface (Schema SolarWall®).

Figure 20 Traditional one-way façade collector. Maximum power 3.5 $kW_{thermal}$.

Figure 21 Vacuum tube collectors as balcony rails: Sunny Woods apartments, Zurich, architect Beat Kaempfen. This is true architectural integration, where the energy unit – in this case, solar hot water tubes – is part of the building itself (Photo Chris Butters).

The picture shows a 1970s variation of a DIY Trombe wall, with a sheet metal absorber, separated from the (un-insulated) rockwall: only works during sunny days. Questionable esthetics.

Purely Active Solar Heating

A long time before the active AND passive solar houses of the 1970s was the beginning of purely active solar heating: there was the MIT solar house, built in 1939, harbinger of a typical post-WWII 1950s approach, a switch of heat source in an otherwise unchanged building concept: a large flat plate collector coupled to a seasonal heat storage, an underground insulated pool, brought to near boiling in the hot

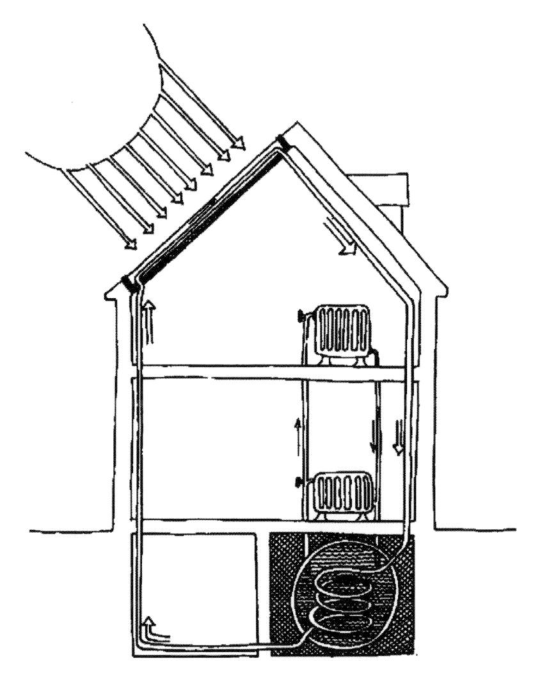

Figure 22 Schematic drawing of the MIT active solar house. From top to bottom: Solar collector, pumped collector circuit, gravity-flow heating circuit (drawing Guy Renaux).

season, and serving as heat source in the cold season. This proved the feasibility of solar heating, but also the high cost of such systems, mainly due to the long-term heat losses of the storage.

Large-Scale Glazed Solar Thermal Plants

Whereas purely active solar houses were quickly identified as impractical and replaced by passive AND active concepts, one concept remained: the large-scale solar heating plant. The basic characteristics of three main contenders for the biggest solar thermal heating plant are shown below.

Table 10 Comparison of collector surface and nominal power of the three largest glazed heating systems. Competition never sleeps: one of the biggest systems is under construction in Qatar, heating water and space for 15 000 students, by a mammoth freestanding collector field of 15 hectares, or 1 m^2 per user. While the local climate avoids heat storage losses (particularly in the sunny Qatar climate), the piping losses and insulation would make this system costly for application in moderate climates, except perhaps for temperature elevation in the cold return side (Source: Solar Heat Worldwide 2008).

Country	M2	MWth
Denmark	18 300	13
Sweden	10 000	7
China	13 000	9

Daylighting

Another interesting application of solar energy technology is the direct use of solar light for daytime lighting purposes: the traditional systems use techniques such as light wells and glazed or unglazed openings with white diffusing interior surfaces (see Figure 27 and Figure 28).

Kleinwächter proposed a lighting circuit circulating a fluorescent liquid, coupled to a trough collector by means of a piping system fitted with an internal reflector. The circulation pump is switched on as soon as there is light demand in the building; a lighting point is switched on by removing the internal reflector on the respective tube section.

> "All too often, the different solar energy systems to be integrated into the building skin are perceived as architectural constraints, following esthetic criteria of the past, while they could be the basis for a new architecture as rich and capable to create emotion as the architecture of the past."
>
> Adrien Fainsilber, May 2011

Figure 23 La Cité des Sciences et de l'Industrie de la Villette (France): "Double skin" for the recovery of thermal energy in the admission of incoming solar pre-heated air between the two glass skins (Source: Adrien Fainsilber).

Figure 24 La Villette (France): "Double skin" on the right, with the pool as a reflector for daylighting and free cooling (Source: Adrien Fainsilber).

Figure 25 Administrative office building (Unedic), Paris (Source: Adrien Fainsilber).

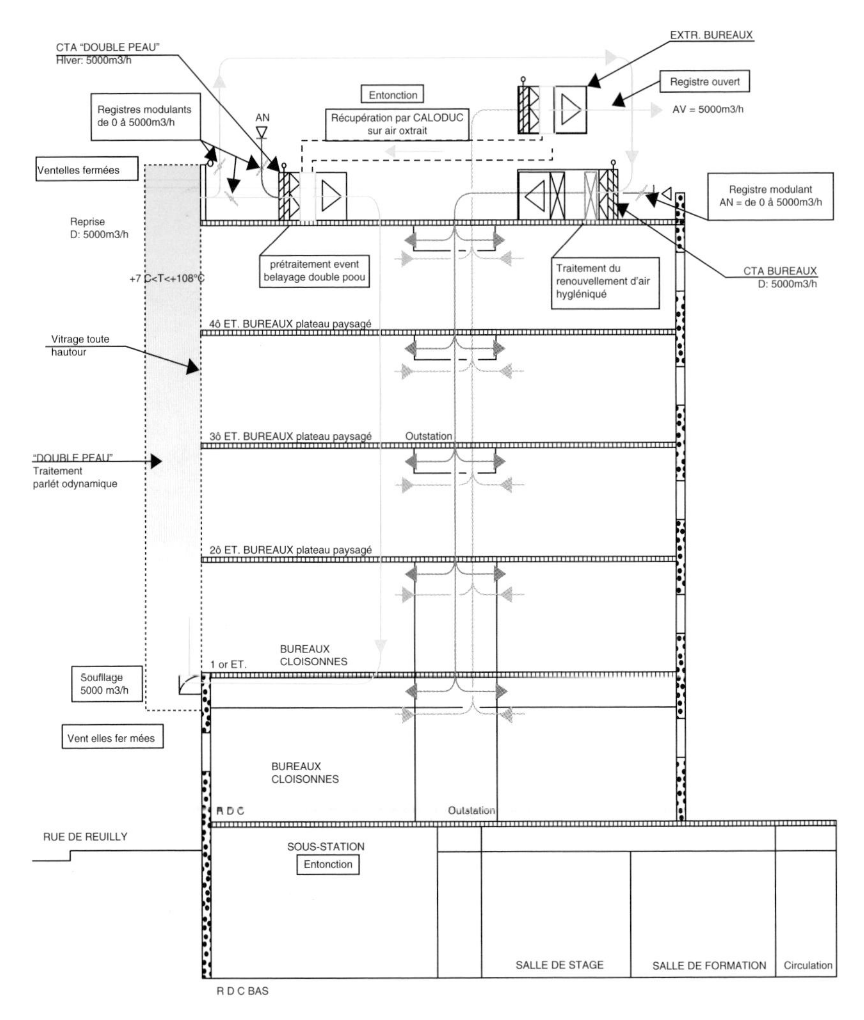

Figure 26 Schema daylighting and cooling (UNEDIC building): the yellow building component denotes the lightwell (transparent light duct reaching over several floors and distributing light via white or refractive diffusion). The red arrows show the forced air intake in the heating mode. The air arrives from the double skin and gets distributed through the offices (Schema Adrien Fainsilber).

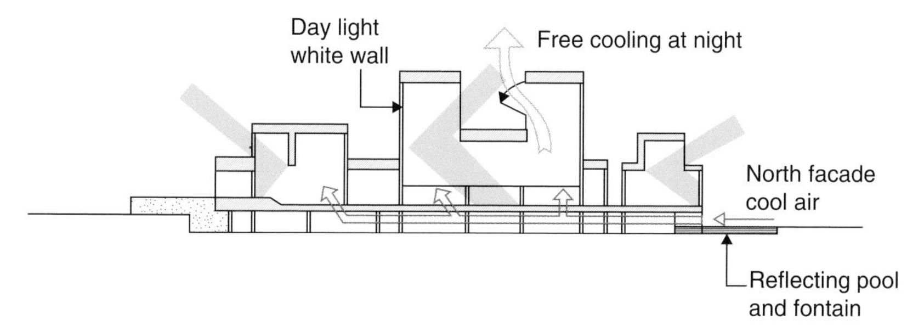

Figure 27 Daylighting and free cooling are two techniques to save on lighting and cooling budgets. Daylighting uses the diffuse reflection on white parts of the building, and reflecting pools and water fountains. Free cooling opens the north facade of the building to the convection driven-cool night air. The high-thermal-mass building components store heat and cold for several hours (Schema Adrien Fainsilber).

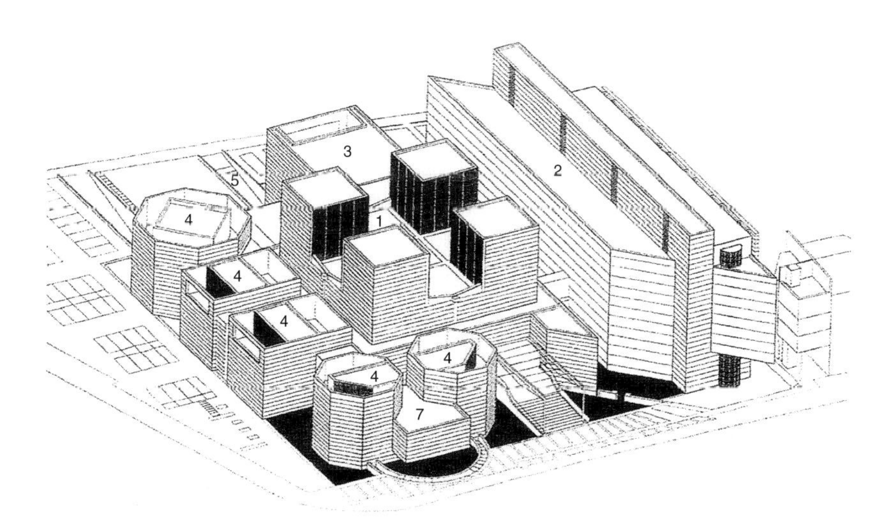

Figure 28 Daylighting and free cooling in the court of justice building in Avignon (France). The massive facades lend an important thermal mass to the building in all seasons. The daylight enters in all rooms indirectly via the terraces. Pools and water fountains contribute to cool the buildings. The warm air accumulated in the high parts of the towers is evacuated through roof openings during the night (free cooling). The ground is cooled in summer and warmed in winter by a thermal water circuit (Drawing Adrien Fainsilber).

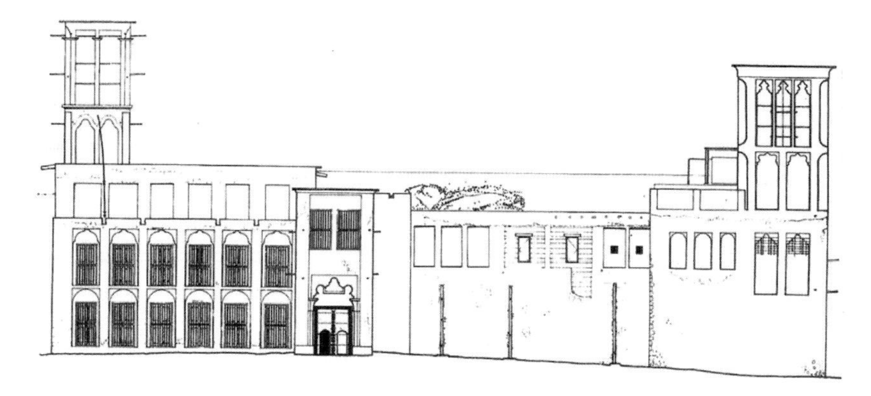

Figure 29 Dubai: For cooling without sun: Bastakia wind towers catch the wind of all directions (Drawing Adrien Fainsilber).

Solar High Temperature Applications

Solar Tower Central Receiver Plants

These "power towers" feature hundreds of precision-mounted reflectors concentrating light on a receiver mounted on a tower, reaching temperatures of over one thousand degrees. Each reflector can be made to illuminate the receiver with a very shallow parabolic geometry at focal distance, up to several hundred meters away. Each reflector is the size of the room of a traditional apartment, and is tracked continuously. It is a chilling experience to witness the focusing process, from amongst the heliostat control mechanism level, on the ground in front of the receiver tower, while the heliostat focus is being pre-set by the operator. During this critical process, the operator illuminates an air volume in space to avoid wear to the receiver; this volume becomes virtually bristling with concentrated light (creating a volume of empty sparkling space, and the insects trying to cross this shiny trap). ***The alert reader finds himself hoping that the operator does not give in to destructive tendencies.***

Figure 30 Heliostat field and receiver tower at the Plataforma Solar de Almeria (PSA) Spain.

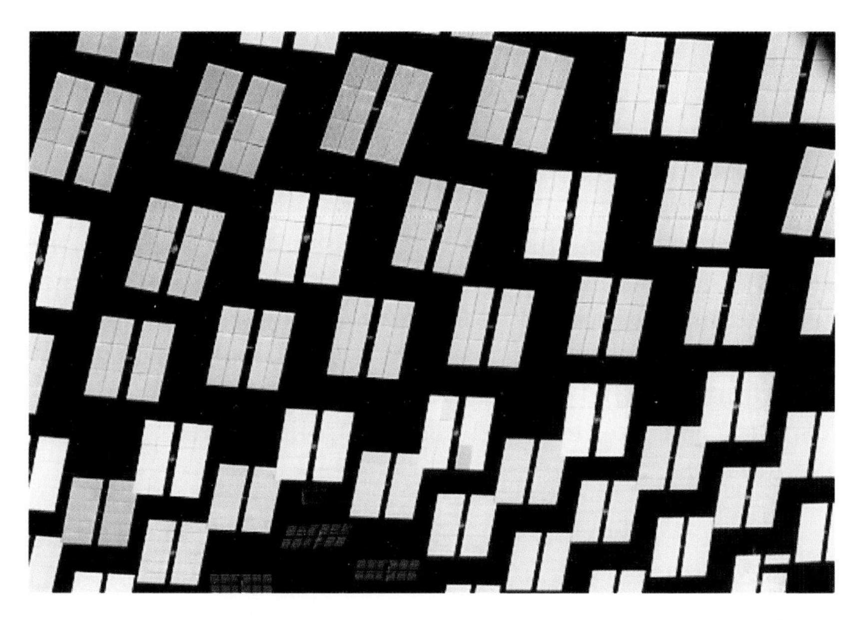

Figure 31 Part of the heliostat field of the tower at the PSA, photo taken from the top of the tower.

Figure 32 The secondary reflector tower of Odeillo, France (Source Wikipedia).

Trough Plants

Trough plants are a more forgiving high precision and efficiency of cylindro-parabolic modules, in north–south orientation with 2-axis tracking illuminate vacuum tube protected absorber tubes. High modularity of the reflectors (of less demanding precision compared to towers) promise cost reduction, higher specific mass imposed by the fact that the whole heat transfer takes place inside of high-tech tubing (compared to in thin air for the power tower). The output heat is transferred to a fluid, isolated from the environment. This opens a potential for trough plants to be used in clean applications (e.g., controlled environment for chemistry).

Dish Stirling

Dish Stirling modules are intermediate diameter lightweight parabolic reflectors where the Stirling engine is heated by concentrated sunlight. The 2-axis tracking requirements are

Figure 33 Dish Stirling engine. The incoming light gets concentrated by the high precision parabolic reflector onto the receiver (hot air heats an air-helium heat exchanger) (Source: fr.wikipedia.org).

not very precise. A prototype series long-term test of these units has been going on for 20 years at the PSA. Results were positive (even with the rather mercurial helium as working fluid).

Solar Chimney Power Plants (CSP)

A potentially interesting concept for solar thermal electricity production is the convection tower: a lightweight tower, of up to several hundred meters height, taking in air which is solar heated in a large greenhouse at the bottom of the structure, driving a low-temperature air turbine for electricity production. The main resource needed is vast desert space, which is not in short supply. Tendency growing; the main enemy is the wind. The first operational prototype in Manzanares (Spain) outlasted its predicted useful life, but only for a short time. The tower was 195 m in height, with a diameter of 10 meters. The collector diameter was 244 m and the collecting surface 47000 m^2. The efficiency of the system was 1%, the efficiency of a version with thermal storage 0.3%. Three (of many) options for improvement of the concept could have been learned:

- Start up small, learn gradually,
- like in sailboats, reduce sail and leave only structure (masts and booms) for when it starts blowing, and
- find more generous investors and go for heavy-duty solutions, such as the gargantuan, highly successful road bridge in Millau (France), built by the same firm who built the Manzanares tower.

However, these will not lower the cost of the investment, which is already prohibitive.

CSP in South Africa (MB)

Most areas in South Africa average more than 2 500 hours of sunshine a year, and average solar-radiation levels range between 4,5 kWh/m^2 and 6,5 kWh/m^2 in one day. The annual 24-hour global solar radiation average is about 220 W/m^2 for South Africa, compared with about 150 W/m^2 for parts of the United States, and about 100 W/m^2 for Europe and the United Kingdom. This makes South Africa one of the world's best climates for solar power, and a number of activities are underway to utilize the resource.

The South African Bulk Renewable Energy Generation (SABRE-Gen) program was initiated in 1998 by South Africa's electrical utility, Eskom. The objective was to enable the evaluation of multi-MW, grid-connected generation systems, to determine whether they could provide viable solutions to South Africa's future electricity needs. Eskom

Figure 34 The first solar chimney power plant was built and commissioned in 1983, in Manzanares, Spain (Source: www.sterlingengines.org.uk). The project was funded by a research grant awarded by the German Federal Ministry for Research and Technology. This solar chimney was in operation for approximately seven years. The successful operation of this pilot plant led to the construction of two small-scale demonstration plants in Sri Lanka. Further work is underway in South Africa.

undertook a prefeasibility study on concentrated solar power (CSP) technologies, and a screening process identified two technologies, solar trough and central receiver technologies, as possible near-term options to be evaluated further. A prefeasibility study for a 100 MW solar power plant with 4000 to 5000 heliostat mirrors, each having an area of 140 m^2, was completed, with a site near Upington in the Northern Cape Province used as a reference site. The prefeasibility study included the compilation of a typical meteorological year (TMY) data file for the reference site, as well as the conducting of a strategic environmental assessment (SEA) for the Northern Cape Province of South Africa, as the most suitable location for possible CSP plants. A large-scale solar chimney plant was considered as the second possible CSP technology. Annual simulation models were developed to predict the performance and costs of the two CSP technologies.

Currently, the research work around solar chimneys is carried out by Stellenbosch University's Department of Mechanical and Mechatronic Engineering, which has been conducting postgraduate research for the past ten years into the performance of large-scale solar chimney power plants. Most of the work has focused on developing computer models that, for example, have predicted that a solar chimney power plant will be able to run 24 hours a day. The advantage of the technology is that enough thermal energy can be captured and slowly released during the day to also store it and then release it in the evening. The biggest challenge, however, is to build a solar chimney power plant, as a fairly high, 1 000 m to 1 600 m, chimney has to be constructed. The University is currently carrying out research by a group of civil engineers, who are investigating suitable construction techniques in cooperation with researchers in the Netherlands and Germany.

Currently, there are no firm plans to specifically build such a plant in South Africa, but the possibility will be

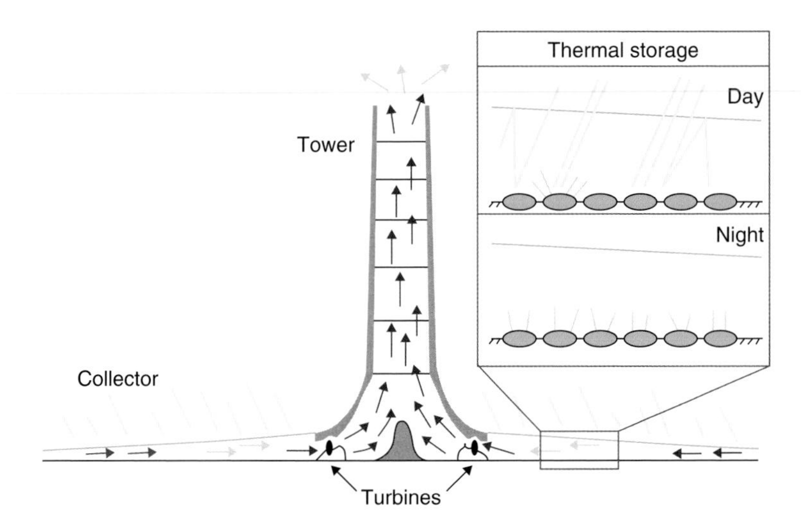

Figure 35 Schematic representation of a solar updraft tower (Source: thenakedscientist.com).

investigated during a process of policy review in the Northern Cape Province, when a renewable energy strategy will formulated during 2012. Investigations are also underway about possible plants in Namibia and Botswana. However, in April 2011, it was announced that Eskom, with the help of a World Bank loan of US $21.8 billion, will construct a 5 000 MW solar park in the Northern Cape Province, as part of an aggressive push to grow its highly industrialized economy and reduce poverty without increasing its carbon footprint. A 100 MW concentrated solar power plant, estimated to cost about $1 billion, is planned for the same region.

Solar Thermal Pumps

The heart of these surface pumps, as demonstrated by Bomin Solar, is an extremely low-rev, flat piston solar Stirling engine, with the piston, heat exchanger and the absorber being the same element. Working fluid is air.

The efficiency of this slow pump comes close to the Carnot limit. However, this limit suffers from the small temperature difference. The resulting machinery is sturdy, but heavy for the small power rating, compared to a PV solar electric pump. Again, this pump can be built with the simplest of means, which confers a high degree of autonomy, a typical representative of "post-big-crisis" technology, with well-stocked junkyards.

Sweeping statement: Mad Max technology, directly inspired by the desert where the Mad Max movies were shot.

Divers Applications

- One of the potentially interesting applications for solar (thermal or PV) energy is the medium- to high-temperature incineration of toxic waste. If strict regulations are enforced, this application could allow environmentally responsible and economically feasible treatment of toxic waste in desert situations (e.g., for refineries).

Solar energy in charge of crude oil extraction, says the alert reader: some acceptance problems might lie ahead ...

- High radiation density material testing: e.g., for the European space shuttle heat shield, which is the critical reentry component.

Cookers

The story of solar cookers could easily fill a book on its own: no other solar technology has stirred up so many passions, resulted in hundreds of designs of different quality, local production projects, field tests, and deceptions, but also success stories of men and women deeply involved in their dream of the sun's heat replacing dangerous, expensive, polluting fuels, sometimes causing barely imaginable levels of deforestation. The situation becomes more dramatic by the low efficiency of the cooking devices (causing, for the popular three-stone fire, almost 10 times the necessary fuel consumption at optimum conditions).

There are two main types of solar cookers:

Domestic Solar Cookers

- Box type (glazed and insulated boxes, with internal absorbers, internal and external reflectors boosters)

Figure 36 ULOG box type solar cookers: this design can be produced locally, with simple carpentry tools.

Conductive solar box cooker T16

Figure 37 A high performance conductive box cooker.

Concentrators, More or Less Parabolic in Shape

Figure 38 SK1200: A deep-focus concentrator (Concept EG-Solar).

Hot Pot SHE

Figure 39 Panel cooker, one of the different hybrids between boxes and concentrators (Photo: SHE).

Institutional Solar Cookers

Flat Plate Collector Cookers (CSC, Schwarzer).

Figure 40 This cooker is one of the few models permitting the user to cook in the shade of the building's insides, which is a serious handling advantage (Photo C. Schwarzer).

Figure 41 The collector corresponding to the reduced size model of Schwarzer's institutional-size flat plate collector cooker, with oil as heat transfer fluid (Photo C. Schwarzer).

CSC Flat Plate Cooker

Figure 42 Solar kitchen in Yavello (Ethiopia). In the foreground, Injera cooker: one recognizes heat pipe vacuum tubes collectors, heating baking plates situated on the inside of the kitchen wall. In the background, three CSC flat plate cooker with 35 liters of pot volume each.

Scheffler Cookers

The biggest and most successful of all institutional solar cookers uses fix-focus reflectors and back-up fuel, and feeds several thousand people daily in India (see Figures 43 and 44).

Most of these types have been tested in two international tests at PSA in 1993 and 1994. The differences in thermal performance between types were considerable, but in the end, in the hot Tabernas desert, most models reached boiling temperature; after different heatup times, all eventually "boiled an egg" (which was not asking much, but such was the challenge thrown in public at solar cooking pioneer Roger Bernard, who picked up the glove and invented the panel cooker).

A question remained: which of the cooker models would be best accepted (in terms of use rate)?

Figures 43 and 44 The steam produced by several arrays of concentrators is transported through steel pipes into the kitchen, then the food is cooked by the steam in huge tanks (Source: Solare Brücke).

A comparative field test, featuring seven different solar cookers models, was organized in South Africa by the Department of Minerals and Energy (DME), GTZ, and German Ministry of Cooperation. Results were surprising: according to questionnaire feedback, all models showed more or less identical use figures; also, the users treated the cookers well after they had acquired them for their own money.

The alert reader could conclude that critical parameters were less preference for technical and handling aspects, but seem more to stem from different "willingness to please" on the part of the users.

Convective Solar Cooker (CSC)

Figure 45 A Convective Solar Cooker (CSC) in Kadugli, Sudan. This cooker, before it was destroyed in the civil war, was used to cook in seven massive pots for 300 schoolgirls. It featured a cold-water storage, a hand pump for the water supply, a solar water heater, hot and cold water on tap, and a work surface protected against sun and rain. It was installed under rough circumstances during a particularly hot Summer Ramadan month by a team of two Europeans (who did not follow Ramadan rules) and two local cooks (who did). None of us will forget.

Autoclave Sterilizers

The heart of these instruments is an autoclave (a heated pressure vessel with a pressure limiter valve), bleeding off steam once thermal equilibrium and absence of air are obtained. Sterilization conditions can thus be obtained and maintained by simple means. Temperature depends only on the pressure, which is being kept constant by the pressure valve, controlled by a manometer, and set at 2 bar or 121°C. To add an independent crosscheck whether the surgical material has been effectively sterilized, a one-way indicator can be used. This is good practice in critical cases and conditions, such as bone surgery in post-battlefield conditions.

Also, sterilization is not only a story of pressures and temperatures, but a complex process implicating technical and communication issues as well as conflicting interests between stakeholders.

In a project, the "northern" headquarters were interested (cost of one-way material), national headquarters were hostile (one-way material being safer and easier to manage compared to locally sterilized material), and field workers were moderately positive (higher local content and more local autonomy).

Another example shows the difficulties in the understanding of the sterilizers, which appeared during a training session: The technician was requested to pour a determined amount of demineralized water into the autoclave. When the result was checked, it was found that he had opened one hundred individually packed units. On a purely formal level, the task was performed perfectly well, since the necessary quantity of demineralized and sterilized water ended up safely in the autoclave. The only problem was the fact that sterilized and individually packed water units are much more impractical to put into the autoclave; they also risk contamination and are radically more expensive.

Sweeping statement by an International Red Cross official: "a solution looking for a problem."

Perceived problem to be solved at the outset of project: sterilization of reusable medical instruments is expensive, and depends critically on fuel supply.

Technical experiences: on a technical level, no serious problems were experienced, with the exception of a steam leak in one of the earlier models: this shows the necessity of indicators. The heating up of the thermal mass, consisting of a small amount of water, the autoclave, and the surgical instruments or dressing material, did not pose any problems. The use of three well-designed booster reflectors allows the use of a simple basic solar box cooker design.

An even more economical approach used a conductive box design (cost perspective US$200 for a conductive box cooker).

Related Experiences

The technical problems were often overestimated in the beginning. "Jurassic" approaches proposed for field tests included:

- a 20′ container fitted with vacuum tubes
- A 10 m^2 FIX-FOCUS concentrator heating up a steel block to 400°C.

Figure 46 The collector for the steam production for the Kadugli sterilizer is placed outside, whereas the sterilizer unit is mounted indoors.

This kind of technical overkill leaves even our alert reader speechless.

More modest approaches included:

- An insulated autoclave fitted with a "boosted" flat plate collector via an optimized two-way air circuit
- An insulated autoclave situated inside, heated by a flat plate collector via a heat pipe (see picture from Kadougli/Soudan).

Direct UV Pasteurizers

UV pasteurizers are arguably the simplest and lowest cost of all solar applications: a simple plastic bottle is filled with water to be sanitized (eventually after filtration) and, one next to another, exposed to the sun. The double action of direct solar UV and heat-up to pasteurization temperature causes the sanitizing action, which is not controlled. This is an extreme case of low equipment cost, but it needs a highly qualified and costly operator: this is probably not the optimum solution. A more adapted solution could be controlled pasteurizing in a solar box cooker fitted with a multiple-use indicator, such as the system developed by Solar Cookers International (SCI).

Example for a direct UV pasteurizer: the Sodis Sanitizing system (Source: SODIS).

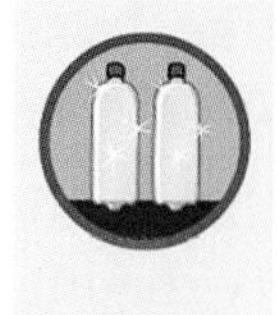

Use clean PET bottles

Fill bottles with water, and close the cap

Expose bottles to direct sunlight for at least 6 hours (or for two days under very cloudy conditions)

Store water in the SODIS bottles

Drink SODIS water directly from the bottles, or from clean cups

Solar Stills for Lead-acid Battery Maintenance

Figure 47 For regular maintenance of batteries for safari vehicles, small solar stills can be top choice. The figure shows an operational prototype of a high efficiency still – not a drop is wasted – or so it seems.

Solar Driers

Solar driers exist in all sizes, for households and agriculture, and are available in different models: direct and indirect, convective and ventilated. They can behave like divas or like saints: divine surprise.

There was a model from Senegal without glazing, easy to produce locally, which delivered the best quality drying results, particularly in the case of the notoriously difficult-to-dry tomatoes, as well as other products.

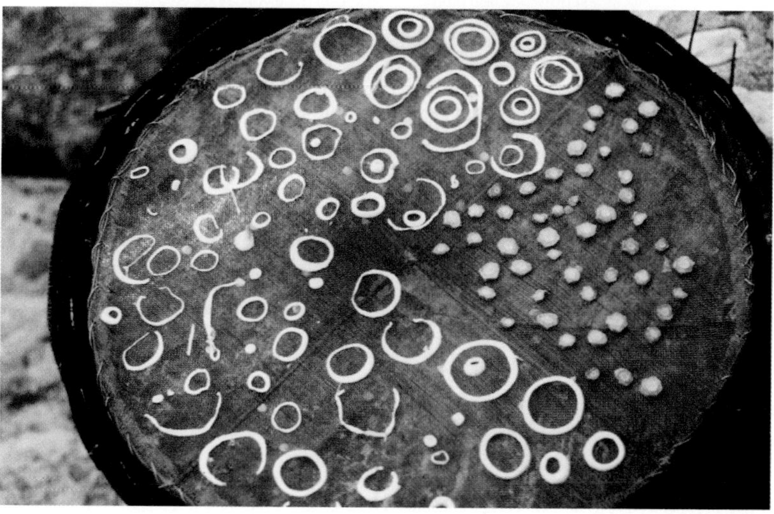

Figures 48 and 49 Different vegetables on trays of the Senegal "lentil" solar drier.

A DIY Direct/Indirect Dryer

Figure 50 The air is heated directly in the front part of the dryer and is then transferred by natural convection to the products.

Lentil Solar Drier

Figure 51 "Lentil": The unglazed dryer from Senegal.

Agricultural Driers

The following photographs have been taken on the occasion of the International Solar Drier Comparative Test in Tabernas, Spain.

Figure 52 Gravitational convective tunnel drier by IST Germany. The chimney, situated on the right, extracts the humid air by natural convection.

Figure 53 Forced convective solar dryer by the University of Hohenheim. The forced convection is driven by a solar PV array of 3 computers fans.

Sludge Driers

Sludge is the "negative value product" of wastewater treatment once the sanitized water is removed. The lower the water contents, the better for future use.

IST (Germany) has developed a flat bed sludge drier, where a small tractor returns the drying sludge in a solar heated drying tunnel.

Solar Thermal Energy – The "Software"

This chapter discusses selected aspects of solar energy that are not directly related to equipment (the "hardware"), but to the use of the equipment: from sizing, testing and evaluation of hardware, to user acceptance, financing, marketing, after-sales service, public support, impact analysis, and so forth. These issues are labeled "software" to avoid confusion with the terminology of the rest of the book. Again, as in the rest of the book, the presentation is not aimed at a complete representative description, but rather at a description in the way of case studies.

Impacts

First of all and foremost, there is the question of the cost impact due to savings in fuel and emission reductions.

Also, the willingness to buy is dependent on different parameters, such as the risk of investment lost in case of the equipment being short-lived or the performance being insufficient, particularly if the equipment is not well known to the public. Then, there is always a risk of unadapted capacity: the equipment can either be too small (this means that the whole operation is not worth the trouble) or too big (which creates unnecessary cost).

An important aspect for the decision to acquire solar energy and plants is the impact on fuel savings and on emissions, resulting (hopefully) in "willingness to pay," the stuff of which market success is made. It is easier to create if the appliance finances itself rapidly and without undue risks, which can be guaranteed by easy exchange, including modularity and compatibility. It should be noted that optimum size can be defined in reference to different bottom lines (economic, ecologic, image-related, or a mix of these). In all cases, it is important to be able to learn from experience, and to change appliance size and capacity by adding or decommissioning modules, which must be on stock and compatible to additional models when needed.

The Market

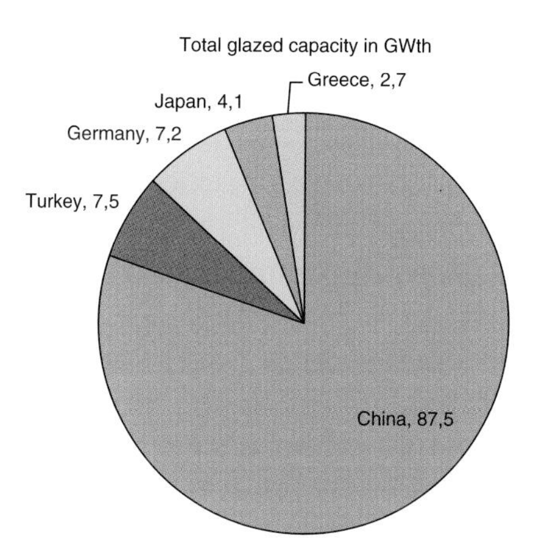

Figure 54 The installed market for glazed solar collectors (top) is dominated by China, whereas smaller countries have higher per capita markets (bottom), indicating an important untapped global market potential, by roughly one order of magnitude.

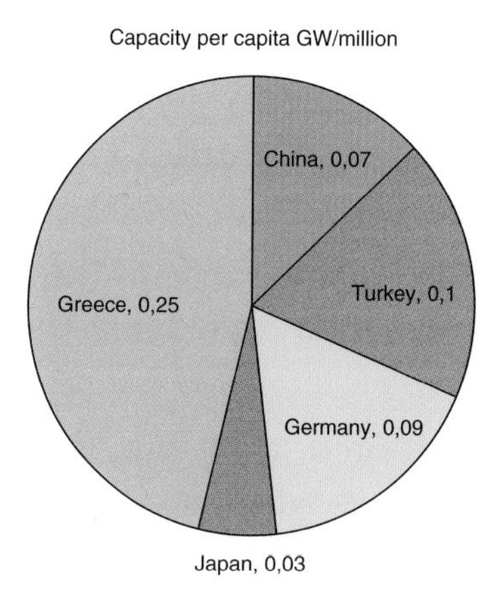

Figure 55 The marked differences between the SWH types on the market in different countries indicates a relative indifference of the market players concerning what is happening abroad: that they seem to be struck by the "not developed here" syndrome would be the alert reader's first guess.

Flat Plate, Evacuated Tube and Unglazed Collectors

The total solar flat plate and vacuum tube capacity in operation by country is given in Figure 54.

Example

Investment usually starts with some hard questions on Investment Methodologies – the solar cooking example from the potential investor:

- What is the edge of the product to be marketed compared to competing products already available on the market?
- How was this edge established?

For renewable energy appliances, an important part of investment credibility is linked to credible monitoring and evaluation of its impact, either in terms of fuel-cost or climate change emission reduction (e.g., CDM) purposes.

Credibility of energy bills usually means metering, and metering can be complex in complex situations such as replacement of traditional firewood use for cooking, which is one of the decisive potential applications for renewable energy. ***The alert reader will ask the key question: how much traditional fuel is REALLY replaced by renewables, or how much food ACTUALLY gets cooked by renewables and how often. The UN Climate Change Commission is busy finding out.***

One of the most promising tracks is the SOLAR COOKER USE METER (SUM). It is based on the linear collector equation.

We had already noted that the linear equation (see page 37) can have a two-way use: characterization of collectors and, if these parametric figures are known, calculation of thermal mass from the evaluation of collector temperature.

One of the most interesting applications is the actual reliable metering of solar cooker use, in order to quantify fuel savings and CDM relevant data, which can serve as a basis for responsible investor decisions.

The Determination of Solar Food Mass and Cooking Time

In order to determine the quantity of food in the pot of the solar cooker, the temperature rise ($T_{final} - T_{initial}$) can be related to the thermal energy input E_{th} to obtain the « thermal mass » (M_{th}) of the food in the pot, defined as mass times specific heat:

$$E_{th} / (T_{final} - T_{initial}) \tag{4.1}$$

The instantaneous thermal power $P_{th}(T')$ can be calculated by:

$$P_{th}(T') = A * I * \eta(T') \tag{4.2}$$

With $T' = (T_{pot} - T_{amb})/I$, T_{pot} the pot content temperature, T_{amb} ambient temperature, A the cooker aperture, I the irradiance, and $\eta(T')$ the linear collector efficiency:

$$\eta(T') = \eta_o - K * T' \tag{4.3}$$

P_{th} can be determined by (2) and (3), and integrated to yield M_{th} by (1). The thermal mass is calculated for each one-minute interval, averages are η_o being the efficiency at ambient temperature, K the linear loss coefficient. Both η_o and K are specifics of cooker models and can be measured directly with the use meter heating up a known quantity of water. Note that η_o and K are constants for each solar cooker model; they do not have to be measured for each individual cooker.

Results

Heat-up data are taken in the temperature range between 50 and 70°C. Unrealistically small or large values can be discarded.

A correction is made for the thermal mass of cooker parts, in particular the pot. Typical values of pot thermal mass are situated between 0.1 litre water eq (for small steel pots) and 0.3 litre water eq (for big Al pots).

To obtain the number of meal portions (MP) corresponding to the resulting M_{th}, the average thermal mass per MP is used, under the assumption that the main part of thermal mass in water-based food is water. A field visit in the use region can determine the local average M_{th} per MP by studying the food volume per capita and meal. The default value is 0.5 l water eq per MP.

In order to determine whether the food has actually been cooked, determine the time that the food has been held above a minimum cooking temperature. The default value can be checked; it should be typically in the order of 70°C and can be compared with typical conventional cooking times and temperatures.

Use Meter Output Verification

To verify SUM results, output data can be compared with actual conditions. The SUM was switched on, then the cooker was exposed to the sun and loaded with a given quantity of water (controlled by an electronic scale). Single and multiple heat-up runs were recorded, with different loads and different tracking intervals, and under different environmental conditions.

Results (in terms of M_{th}, cooking time, number of successful cooking cycles) correspond to the actual cooking history.

Thermal masses were 3 litres, 2 litres, and 1 litre for the first, second, and third heat-up cycle, respectively. Note that the calculated thermal masses correspond to water and pot (here, the pot thermal mass is 0.3 litre water eq.). With this correction, the precision in the determination of thermal mass for the K10 cooker is in the order of 10%.

The alert reader, as always a little on the nagging side, is asking (again...) whether all of this is worth the trouble. The answer is clearly yes: remember what the question was; it was NOT a rigorous quantitative determination of fuel savings, but, first of all, a credible answer of whether solar cookers are used at all.

Comparison of Simulation with Actual Cooking Action

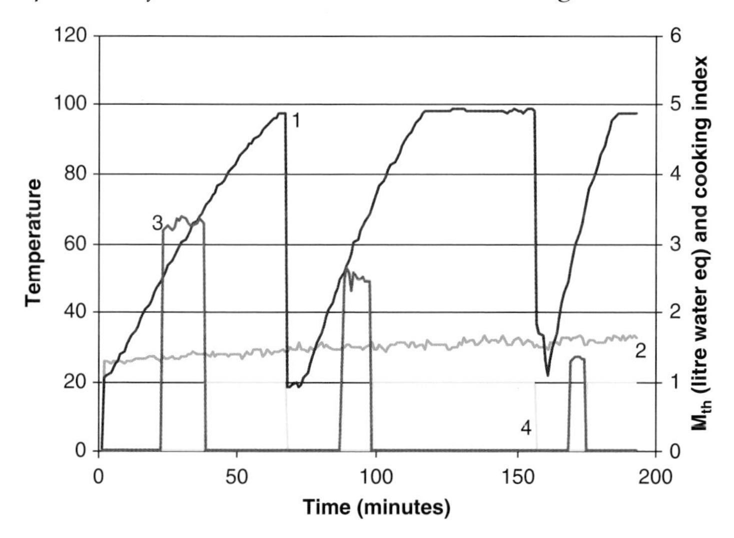

Figure 56 SUM results for a K10 concentrating solar cooker.

The table presents SUM results for a K10 concentrating solar cooker: on the basis of measured pot bottom temperatures (1), irradiance, and ambient temperature (2), SUM calculates instantaneous thermal mass (3) and cooking index ((4) data points: 1 indicates cooking, 0 no cooking). The cooking time count is stopped when the temperature falls below an adjustable level (set here at 70°C).

Table 11/Figure 57 On the left, in the yellow column, the input parameters are shown. On the right, the corresponding output parameters. As an example, the respective gross thermal mass has been calculated as 6,13; the food was cooking during 36 minutes (0,6 hours).

Input Parameters		Meter Output	
Cooker data		irradiance anomaly?	
Cooker aperture m^2			0
	1.54	Total gross thermal mass	
Cooker optical efficiency			6.13
	0.32	total net therm mass	
Cooker loss coeff W/m^2K			5.53
	1.11	Total solar hup Wh:	
T start "heat-up":			549
	50	hours cooking:	
T start "cooking"			0.60
	70	number of cycles:	
GHG (CO_2eq)			2
wood 3 stone		number mealportions:	
3 Stone fire efficiency			9.868
	0.1		9.868
non-sustainable wood			
	0.5	CO_2eq	

Input Parameters		Meter Output	
kg CO_2/mp		wood 3 stone non-sustainable	
	0.8	kg CO_2/mp	
kg nonCO_2/mp			0.6
	0.2	kg CO_2 saved	
Total			5.9
	0.6	Carbon credit $	
			0.12
carbon value $ per t CO_2			
	20		
net heat input / MP in MJ			
	1.21		
M_{th} minimum for average:			
	0		
M_{th} maximum for average:			
	6		
M_{th} empty:			
	0.3		
M_{th}/mp			
	0.5		
step minutes			
	1		

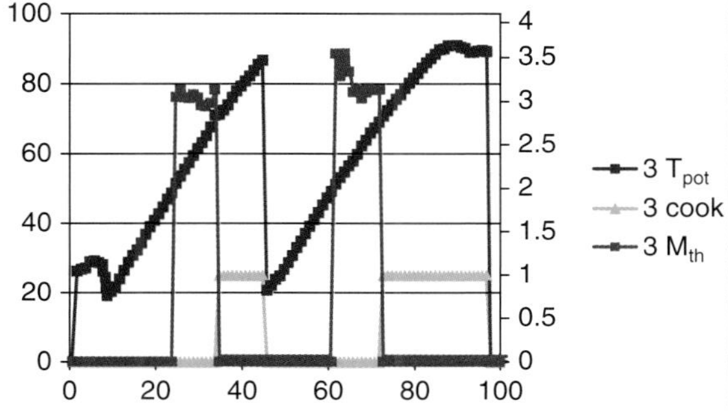

Clearly, on term of absolute precision, such a sensor cannot match a pyranometer.

The table presents the details of the corresponding input and output parameters.

5

Solar PV

Solar PV is a happy partnership of electronics (for the solid state electricity production) and electricity (for the distribution, use, and transformation of this electricity). In contrast to solar thermal electricity, PV does not rely on the Carnot cycle and is not dependent on wasteful temperature differences, although PV cells will also build up temperature differences, but these temperature differences are not needed by the cell to function, quite on the contrary: the lower the temperatures involved, the higher the cell efficiency (see below).

PV – Basic Characteristics

In the following section, we will present some basic characteristics of solar photovoltaic (PV) cells, the most basic of these characteristics being that there are not many of them, except that PV cells are "just" producing electricity and a bit of heat, period. No noise, no smell (if you evacuate the waste heat, which is also recommended for reasons of efficiency and safety, see below), no after-sales service (except for cleaning of the glazing, if you follow the *lege artis* rules). Of course, if you do not, it is at your own risk...

On a closer look, there *are* some more characteristics, but nothing needing ultra-sophisticated technology, just good old electricity: a dream for an installer, and a nightmare for competing energy solutions, or peripherals such as inverters, batteries, and charge controllers, whose earthly weaknesses stick out in comparison to PV cells like sore thumbs. ***Also, as the alert reader says, a nightmare for authors of solar PV books who are afraid to lose the last remaining drama potential of their trade.***

Shading

One more such minor characteristic of PV cells, which influences different cell types at different degrees, is a more or less sharp drop in output power on partial shading of the cell or cell array. In this case, the shaded zone short-circuits the electricity produced in the sunny zone, before it can do any good to the prospective user or to the bank account of the investor who financed the PV power station.

The solution is simple: separate the cells into zones and link up the zones via diodes. This also introduces a further error potential – ***diodes are directional, which is their raison d'être, thinks the alert reader, who starts to feel a bit sleepy.***

Thus, care must be taken to separate "strings" (PV cells in series) in order to avoid internal short-circuits due to shading, e.g., by walls, chimneys, trees, or masts, sails, or other slow- or fast-moving light blockers ***(someone will end up calling them "Dark Shadors," thinks the alert reader, who is under the impression that the shading issue is somewhat overrated).***

The Temperature Effect

Many people have been disappointed by the power drop shown by PV with rising temperature: just when the sun shines brightest, when expectations are highest, *and when cold drinks should be, well, coldest, the efficiency of PV is at its lowest.* Too bad, but good for something: due to this temperature effect, simple micro-sized PV cells can be used as low-cost irradiance sensors. Wired in short circuit, the current, which is uncritical to measure on small cells, is proportional to the incoming light intensity. On absolute precision terms, this will not replace a pyranometer.

Electricity and Grids

Electricity and grids are part of the potentially most polyvalent distribution and virtual energy storage tools available today, but let us avoid false conclusions: electricity and grids also have disadvantages, some of them directly related to their qualities that could be termed "disqualities," such as:

- Universality: one plug does it all, from space heating, to air conditioning, water heating, entertainment, communication, and lighting, which creates an attraction for the corresponding technology solution, even if more economical solutions with higher local content and lower (accident and other) risks exist.
- This universality implies the bet that the grid will always be able to cover any load state, just by adding or removing the necessary capacity, on time, at virtually any price and user behavior. In extreme cases, this can lead to overload, but also "underload" system states with negative electricity prices (see Introduction).

- Convenience: no handling of fuel or combustion residues (these byproducts, including their environmental impact, have to be managed by the operator of the power plant).
- Wide variety of transport and branch out capacities, from GW down to mW power rating and low transformation cost.
- Self-perpetrating potential: just like roads or pipelines, once a grid is installed, the decision to change a distributing system is increasingly hard to take, because of the defensive competition the grid opposes to new solutions.
- A strong "fata morgana" (the optical illusion of surface water which has misled many desert travelers) attraction: once a grid has entered into the realm of remote possibilities, other options face a tough uphill drive. Who wants to settle for an "ersatz grid" if the real thing could be available soon?

The main point on the negative side is the upfront cost for the introduction of the grid, in particular for the exceptionally high initial investment needed, which is due in advance for the power plant and its specific infrastructures. In a society whose members expect that major investments are the responsibility of "the state," this leaves utilities with the option to invest into the grid, in priority for high population density, high-income clients in the direct neighborhood of grid points. The main problem is that potential clients outside of these categories have low priority, and will have to wait off grid, or go for other options, such as solar. ***The alert reader will have noticed that these "other options" do not necessarily bring electricity to places where it is critical: a) to get it; and b) to get it in an economically and ecologically sustainable way. Otherwise, it will be another way to further improve access to electricity (and to all its comforts)***

for those who already have it, and whose "willingness to share" will be remembered and taken as a model by those who do not.

In general, PV can be used grid-connected or off-grid. With grid-connection, all produced energy (depending on weather conditions) is fed into the grid, with the need of taking into consideration all regulations and problems of grid management, and balancing of supply and demand. The end-user will take from the grid in case of demand. In this sense, the grid serves as "virtual storage."

Off-grid, PV can either be used directly when generated (e.g., solar pumps), or in combination with electric storages (e.g., batteries). In this case, charge or load controllers are necessary to extend the lifetime of the storage device, and excessive use will shorten the useful life of the storage. Up to 10 off-grid applications could be connected with each other to improve performance and availability (see Vision 3).

PV Applications

The following section shows selected applications of:

- dedicated solar PV technology (each appliance having its own "dedicated" solar power supply), and
- general-purpose supplies, serving for several different appliances and purposes,
- grid coupled and isolated PV,
- for a wide variety of uses, in transport (rail, road, water, and air), architecture, lighting, communication, entertainment, domestic appliances, and more.

Enjoy...

Figure 58 2394 solar panels with a total system power of 221,63 KW are placed on the roof of the Vatican's audience hall. The system produces about 300 kWh/y, which corresponds to the annual needs of over 100 households (Photo Solarworld).

Figure 59 SolarWorld GT: for the first time, a solar car will be driven around the world. At the wheel are engineering students from the University of Bochum, Germany. The lucky team ("problem based learning") will take about one year to circle the globe. The first leg was a 3000 km "pedal-to-the metal" through the Australian desert. No filling station. Top speed 120 km/h, no radars. GO! (Photo: Solarworld).

Figure 60 The "Solar Bullet train" connecting Tucson with Phoenix Arizona, is still in the planning stage. The necessary energy should be produced by overhead panels (Source Solar Bullet). Also, in Belgium, a similar idea was realized in summer 2011: 16000 solar panels on the top of a tunnel produce about 3,5 MW/y. This is the energy needed to run a train at a speed of 300 km/h for approximately 500 h (Source http://www.lefigaro.fr/environnement/2011/06/06/01029). Really? says the alert reader.

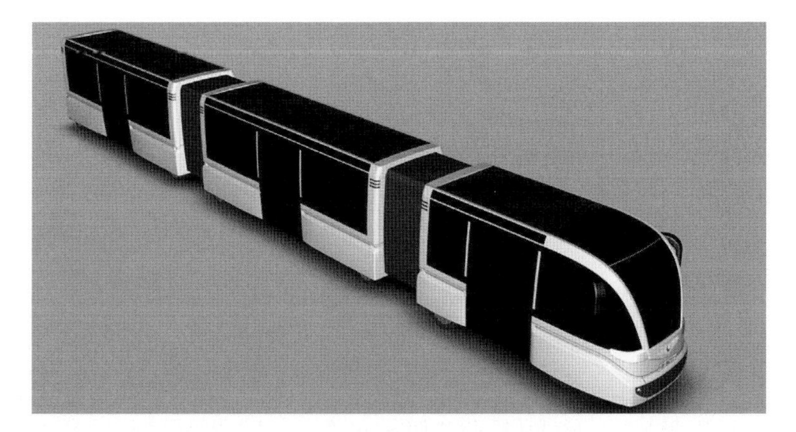

Figure 61 Urban solar train: "where is the PV?" asks the alert reader. The electric locomotive pulls two independent wagons. The panels, which are placed on the roof, produce the energy to operate accessories like air conditioning and video system (Source: Swiss Road Trains).

Solar Air Planes

Solar Impulse, based in Switzerland, entered the 21 century with an extraordinary challenge: to build a solar plane, able to fly day and night without any fuel (nor pollution), just "fueled" by the sun. After seven years of planning, construction, and testing, the first solar day (and night) flight in history, lasting 26 hours, 10 minutes, and 19 seconds, was achieved in 2010. In summer 2011, European flights between Payerne (Switzerland) and Bruxelles, Bruxelles and Paris, as well as Paris and Payerne were operated with the solar plane HB-SIA, proving the huge potential of new technologies for saving energy and applying renewable forms of energy.

Technical datasheet of the solar plane HB-SIA:

Wingspan: 63,40 m

Length: 21,85 m

Height: 6,40 m

Figure 62 Solar Impulse HB-SIA during a flight over Switzerland (photo ©Solar Impulse/Jean Revillard/rezo.ch).

Weight: 1 600 Kg

Motor power: 4 x 10 HP electric engines

Solar cells: 11 628 (10 748 on the wing, 880 on the horizontal stabilizer)

Average flying speed: 70 km/h

Take-off speed: 35 km/h

Maximum cruising altitude: 8 500 m (27 900 ft)

Solar Impulse, in collaboration with component producers and material suppliers, continues the adventure: a new solar plane (HB-SIB), even bigger than its predecessor, is under construction. The aim is to attempt to fly around the world in 2014.

Solar Boats

Figure 63 Solar ferryboat on the Lake of Constance (Source: Deutschland – Land der Ideen).

Figure 64 Vauban, the landlocked solar "ship" by Disch in Freiburg (Photo Anne-Laure Ruffin).

Figure 65 The Solarschiff ("solar ship"), a plus energy center designed by Rolf Disch, features most of the current solar techniques. Commercial activities are situated at street level, with offices on the intermediate floors and penthouses on the top (Photo Chris Butters).

Figure 66 On the "Wohnen und Arbeiten" apartments, an early passive project in Vauban, the photovoltaics double as roof covering over the walkway. Two functions in one will save costs (Photo Chris Butters).

Completely different from the abovementioned solar ferryboat (fig. 63) is the solar catamaran "Tûranor Solarplanet," which started its journey around the world in September 2010, propelled just by solar energy. The boat will cross the Atlantic Ocean, Panama Canal, Pacific Ocean, Indian Ocean, Suez Canal, and the Mediterranean Sea.

Some technical data:

Length: 31 m (with flaps: 35 m)

Module surface: 537 m^2

Average engine consumption: 20 kW

Width: 15 m (with flaps: 23 m)

Installed power: 93,5 kW

(Source Solarplanet)

Dedicated Power Supplies

The prices of PV are falling rapidly, which implies that, for small power supplies, it is becoming increasingly more practical to include separate (dedicated) power supplies for each appliance, although this development is limited by the cost of the necessary control and storage devices.

"Plug" Power Supplies

Power systems serving to supply electricity indirectly via a plug can be termed "plug power supplies."

Traditional contenders to dedicated power supplies are battery chargers, the most common being Solar Home Systems (SHS), which typically consist of a PV cell or array of 30 to 200 Wpeak, a control unit, as well as batteries, that deliver DC electricity for domestic use, mainly lighting, radio, cell phone charger, and television.

Solar battery chargers generally have the following components:

- a PV module support structure with fixed azimuth and elevation position. An optional tracking device can improve the yearly solar "harvest" by up to 40% (single axis compared to an automatically tracked version).
- to replace one way, disposable, dry cell batteries for all low power purposes where voltage stability is uncritical.

Plus points: at times when PV cells were important cost factors for universal battery chargers (such as solar lanterns, which "piggy-back" cell phone battery chargers), these were optimum cost solutions to become independent from dry cell procurement and disposal in isolated sites. They can be integrated into outdoors clothing, backpacks, camel loads, sail covers, tarpaulins, and so forth.

Minus points: Tricky handling, and easy losing, of numerous multiple adapter plugs.

With the recent PV price drop, dedicated power supplies with integrated charge controllers are becoming more and more interesting.

PV Power Plants

After enjoying revived interest, business is sharply dropping, which has the disadvantage of lower efficiency and hence, lower output for a given number of units and other factors, such as the area of land required (low irradiance efficiency).

Also, PV can replace fabric, hence saving costs, and they are very appropriate for individual installations and for evolutionary grids (see visions 3).

Advances in photovoltaic were slow for some decades, until PV became an industry, expanding at a rate of over 20% per annum until the bubble burst. As it becomes more economical, it has some big advantages, not least that it can be applied on any scale, from a single unit on a street light or isolated holiday cottage to a large power plant. And electricity is still often the easiest and cleanest energy carrier.

Here again, great advances are being made as regards integration into buildings; there are now PV roofs where the units replace normal roof sheets, and even roll-on systems, which can be rolled out on top of existing flat roofs on industrial and other buildings. PV arrays have also been applied as balcony rails or semi-transparent roofing. Applications are therefore potentially almost universal; this, for any marketing industry, is a big advantage.

However, once again, we need to remember the big picture. PVs do not produce very much energy per square meter (figure around 150 kWh/m^2.year in France). This means that to supply a normal house with PV energy, one would pretty much need to cover the whole garden with

Figure 67 Designed by Rolf Disch, plus energy houses produce, on a one-year average, more energy than they use. Great success with the public: the "DM" in Freiburg must be one of the most photographed billboards in Germany (Photo Chris Butters).

Figure 68 Thermal/photovoltaic power plant/condo Bennau has been accepted by conservation authorities, and coexists nicely with a protected church. For peak winter demand, a fuel burner takes over. Excess heat is sold to the neighbors, which makes this smart design the first condo plus energy house in Switzerland (Photo Solarmedia).

PVs. But when one advances the energy efficiency agenda, reducing the end use needs to a minimum, such as in extreme low-energy houses, then the picture becomes really interesting; because the south-facing roof area is then enough to supply the house. In rounded figures, in a moderate climate with an average daily sunshine duration of four hours (or 4 kWh/sq.m), a south-inclined roof of "normal" size, say 50 sq.m, will intercept a total of around 73,000 kWh, which corresponds to more than 10 times the total energy needs of the house. Taking into account the typical system efficiencies (5 to 15% for PV, 30% for SWH), this can even leave some surplus, hence the appellation "plus energy houses" (see Figure 67).

In practice, these buildings produce a surplus in the summer, which is sold to the grid (where it might be useful for air conditioning load peaks), and the buildings buy back a little in winter. Here again, we see the need for integrated energy planning, where the individual and large scale solutions can work in synergy; the energy supply system as a whole must be brought into the picture and cater for this. Also, tariff systems must be adapted to send out the right signals to the user.

6

Conclusions Beyond Solar

We have presented examples of the explosive development of solar energy applications, of technical progress, and cost decrease reached. Considering the cost explosion of fossil energy carriers, we can safely predict that the energy market is heading for a critical showdown between:

- On one side, the old system, built on wasteful use of fossil fuel and an unsustainable development model, rich beyond imagination, but unable to satisfy the expectations of the world's poor, while the world's rich *and* poor are waking up to the fact that no amount of money can make the planet bigger than it is.
- On the other side, a new system in the making, not yet a real threat for the old system, but with one strategic advantage, often overlooked in debates on costs, technology, and environmental impact: it is sustainable.

We must also make it desirable, more desirable than another few decades of watching fossil fuels run out. It is not only impossible, but also unnecessary to repeat old errors. So why insist? Why keep mistaking wastefulness for luxury, when there are better, more respectful ways to live? This would create millions of jobs, in all aspects of technological innovation, investment in future infrastructure, while keeping fossil fuels and other resources in the ground, and clearing the dirty rivers and skies, if – and of course there are ifs – the rich accept to renounce things they do not need and probably do not want – and if the poor renounce *fata morgana* development promises in favor of the only real thing: an attractive sustainable technology we all can afford and share.

Case Studies

The following final part of the book presents "case studies" of the future which we will call outlooks. These outlooks:

- try to motivate the reader to formulate her or his own outlooks
- do not try to be representative in any sense
- but try to convey a sense of urgency, like best bets.

"Visions, urgency," next thing they will hear voices, says the alert reader. My job is becoming dangerous. If it was not for all of our friends...

Solar Energy in a High-density Urban Environment

- The growing population density in the mega-cities will limit the access to sunlight per inhabitant more and more. This will put strain on the regulatory systems. Minimum solar efficiency and "irradiance laws" will have to be enforced.

- Further reduction of consumption through energy-saving incentives and measures will make the incoming solar energy supply go a longer way.
- Technical improvements of solar energy devices for sub-optimal settings (e.g., larger acceptance angles, both in elevation and in azimuth) could further increase the usable solar collection area and daily, as well as seasonal, collection time; and this with and without tracking.
- Bionic development methods to increase acceptance angles, based, e.g., on the outer ear geometry of grazing animals, or optimized highly efficient collective plant-growth strategies could be checked for possible solar application in this framework. *Buildings with ears… evolution never stops…what will they come up with next?*

I think I will have to get involved, says the alert reader to himself. There are many solutions nobody talks about: the multiple uses of solar collectors, doubling as windows, as over-heating protection, daylighting, and I do not even like crowded places. Larger parts of the 30m^2 PV needed per household would not have to be installed in situ of the consumption site, and could be produced offsite, on non-domestic buildings and public surfaces. This tendency is already well on its way.

The next section presents a case study-concept for high population density "habitat," in a vision by Adrien Fainsilber.

"Solar Casbah": Low-Cost Solar Energy Vision

The starting point for this concept is radically different from the typical picture of solar energy: it is the Casbah, densely populated urban hills, with heating needs in winter and cooling needs in summer.

Globally, the market introduction of low-cost systems, in the long run, will have a higher impact on energy and resource use, emissions, and the availability of desirable energy services than high-cost systems. The reason is consumption – ready to double, and to double again: higher unmet demand for low-income consumers entering the market usually at a very low energy consumption, but ready to explode as soon as the possibility arises, in terms of availability and affordability:

- The good news is that there is enough sun for all to go around (according to a rough estimate, only about 1% of the world's hot deserts would be enough to cover the entire electricity needs, in higher population density situations.
- The bad news is that the practically available per capita solar input is limited and competes with other uses. Thus, "irradiance-efficient" solutions must be sought to share the place in the sun among all consumers. One possibility is multiple uses of "irradiance space": whereas in a classical house the solar collector can stay where it is, rain or shine, mobile and/or multiple uses can save space in the sun. An example of such multiple-use systems is presented below: the "solar Casbah." This system is not meant to be a unique generalizable solution, but an example of an adaptable "plus energy" (Disch) development approach; several such approaches are presented (see vision on grids below).

It is based on a pre-fabricated modular unit producing a large part of the domestic energy needs by a sunny terrace and façade on the housing unit, with the following functions:

- Thermal storage in a conductive floor slab (following the technique of component activation for buffering of room temperature variations).
- A novel multi-functional solar collector: when the solar collector function is not needed, the collector can be transformed into a window or glazed door, with 3 possible characteristics to choose from: Open or closed door/window, reflecting surface against overheating or mirror for beauty, transparent for daylighting (depending on conditions), plus absorbing, when the collector is needed for heating or "free cooling" purposes.
- Water heating, by a split element natural convection SWH, heating water (automatically or using manual priority controls) with excess solar heat.
- Cooling of the facade and the rooms via the buffer slab, through air intake during the night in the hot season, as soon as the heating-cooling manual controls are positioned in the *"cooling"* position.
- Daylighting: independent of the momentarily chosen collector function, a daylighting device directs a small part of the incoming irradiance via a reflector (e.g., Fresnel or diffuse white) on the white ceiling.
- Lighting by complete equipment or pre-equipment (wiring) for PV-SHS, with standardized, local AC-grid coupling, inverters and two-way electricity meters. In this way, with minor initial investment, electricity can be traded between households, with utilities, and in the framework of CDM carbon projects. ***The time standard for the integration between the local***

> ***grids and between local and central grid can be downloaded via satellite dish which assures additional user acceptance and willingness to buy, at least among soccer and football fans... which can induce a certain need for refrigeration, observes the alert reader.***

It can be noted that such a multipurpose concept is relatively complicated, and needs a good interface in order to be practical to use, which implies some investment in R&D and product development. ***On the other hand, the availability of investment capital has been shown lately to be***

Figure 69 Adrien Fainsilber's vision of the solar Casbah.

facilitated by a recent trend towards gentrification, which has been observed in the city of Marrakesh, where run-down Casbah houses are being sold at excessive prices to the rich. This situation creates its own problems. Some of which might end up being more dramatic than typical solar energy issues. "Pecunia Non Olet," thinks the alert reader, who hopes that some of the willingness to buy will concern renewable energy and will end up serving the poor...

The view of the solar Casbah shows terraces made accessible by multifunctional solar collector devices, which serve as doors and windows when the need arises. Note the daylighting units over the windows and the hot air "free cooling" outlets (curved arrows).

An Up-market, High-Tech Vision

Not all solar visions are concerned about low-cost sharing of available "irradiance space." The following lines are part of a "future vision 2030" by the European Solar Thermal Technology Platform; they see the most urgent developments in the sector of ready-to-install vacuum-insulated elements. ***Solar energy for a minority, says the alert reader, these guys are humming with too much energy already, and not much will be left for those who need it. See for yourself:***

"...Various solar facade and roof modules will be available, for example solar thermal collectors for water or air heating, photovoltaic modules for electricity generation, as well as modules with transparent insulation for directly heating the walls. Facade elements used for heat insulation of existing buildings will be significantly thinner and, at the same time, offer greatly improved insulation characteristics, for example through the use of vacuum insulation. The elements will be offered in a wide range of standard raster sizes and will offer the architect all possibilities for adding full-surface solar facades to the building. The ability to combine solar and opaque elements with any desired surface will extend the architectural design possibilities and offer the chance of providing a complete solar energy solution." (Solar Thermal Vision 2030, ESTTP, 2006).

Solar Thermal vs. Solar PV: The Battle of the Water Heaters

Revolutions have a tendency to devour their own offspring. This is not only true for the renewable vs. fossil energy cost issue, but also for SWH from different solar sources. Recently, the flagship of renewables, the solar thermal water heater, has come under attack as being more costly than the PV water heater, in spite of their widely differing nominal energy efficiency (roughly 80% for thermal, 15% for the PV).

Several reasons are pointed out:

- Lower production costs due to the recent PV price collapse.
- Lower installation cost.
- Lower integration cost into existing systems, including grid electricity.

The issue is far from being settled yet:

- All of these factors are direct consequences of the subsidies for PV, which might be reversible by similar subsidies for solar thermal water heaters.
- Also, all of these costs depend on the selection of system types for the comparison (so far, mainly upmarket). Low-cost thermal SWH are much more affordable than PV SWH, and this is where the bulk of the market is.
- Finally, the low "irradiance efficiency" of PV will limit the applicability of this technology in high-density environments (see below).

The alert reader concludes:

- *This is another example of a multiple bottom line case.*
- *Yet another example of the market volatility of solar energy which could create (positive as well as dangerous) opportunities for all stakeholders.*

On the other hand, thinks the alert reader, this could allow the startup of a PV plus thermal technology, where thermal contributes the advantage of high energy efficiency and low space needs, and photovoltaic the advantage of easy integration into all types of grids (see following section), a whole new total solar technology – let us get carried away...

And do not forget, the alert reader...is you.

The Evolving Grid:

How users (and other stakeholders) can let small grids grow big:

Comparative field tests experience suggests that favorable conditions for the market introduction of domestic solar PV (for lighting, battery charging, entertainment, and similar purposes) include:

- Ownership and technical responsibility in the user's hand have been shown to be the main conditions for success.
- The investment risk is perceived to be acceptable by the user if the resale value is high and the investment is reversible.
- Easy step-up of investment volume and unit power rating by exchange of several small units

for compatible bigger size modules of the same total power rating.

- Linking of compatible installations to form growing small grids or meta-grids, becoming increasingly attractive to users and investors.
- Healthy competition between different grids can lead to mutual "evolutionary take over" without investment loss, e.g., through competitive "under the grid" electrification by small PV units between the inaccessible pylons of the big grid.

Remarks on Energy Planning

The application of solar energy technology implies some requirements concerning city planning: if we are to use solar energy, then streets must be laid out so that most roofs are more or less south-facing, and heights must be zoned so that neighboring buildings do not shade each other. Consideration for solar energy should, therefore, be included in urban design. The EU has recently highlighted the need for attention to this in all future city planning. The municipality of Davis in California was an early example of solar zoning, with the intention of ensuring "solar access" to all properties. In many recent planning concepts highlighting sustainability, one can also see a tendency for the designers to point most buildings towards the sun.

Unfortunately, not all building plots can be perfectly planned, and optimum orientation tends to be the exception to the rule. We have to plan and settle for situations which are not ideally situated in space and time within the acceptance angle of the solar device. The development of building concepts adapted to these requirements will be one of the most important tasks in the field of solar energy, another one being R&D into multiple use solar collectors (see Vision 1).

It seems a likely scenario that in the not-so-distant future, solar energy production will be standard for all roofs, everywhere, to be not only providers of shelter from the weather, that is to say rain shields, but energy producers, including thermal solar panels, but mainly photovoltaic. The planning of national energy supply systems needs to take this into account.

Among other things, this scenario does lessen the advantages of some alternatives, such as biofuels-based district heating systems, since everybody needs a roof anyway.

Also, this trend turns energy policy focus back in the direction of electricity. This would be a significant shift; in recent years the accent has been on getting off electricity, since this was primarily produced in fossil fuel power stations.

In general, it is not useful to evaluate a technology, solar or other, in isolation from its full context of application. The solar age is nevertheless our future. Having good applications and abundant supplies of solar energy can only be an advantage.

Solar for Existing Settlements

It is often stated that most of tomorrow's buildings are already built. This is particularly true in developed countries, where the rate of renewal of the building stock is typically only around 1% per year. Existing settlements and cities thus pose our major energy problem. Solar energy applications have a considerable potential here too. Much of the solution again lies in first reducing demand through energy-saving retrofitting. However, many existing roofs can be solarized with both thermal and PV applications. Some facades will be available too, although many, in towns, will not receive much sun. In the case of fairly low-rise traditional urban buildings, solar roofs may in the future typically provide 50% of the energy needs, once these have been reduced through energy-saving measures.

One of the main difficulties in improving existing settlements lies in their structure, however. One may change or replace individual buildings, but the overall layout to a large extent determines energy use, in particular for transport. It has often been said for example that the typical sprawling American suburb is designed "on the assumption of cheap gasoline forever" (Clark Bullard): it is almost impossible to get around without a car, and this sprawl is typical in many other countries too.

This is obviously an issue that lies well outside the scope of this book but is important to remember. On the other hand, suburbs do present an easy opportunity for individual solar installations.

Also of Interest

Check out these other related titles from Scrivener Publishing

Zero-Waste Engineering, by Rafiqul Islam, ISBN 9780470626047. In this controversial new volume, the author explores the question of zero-waste engineering and how it can be done, efficiently and profitably. *NOW AVAILABLE!*

Sustainable Energy Pricing, by Gary Zatzman, ISBN 9780470901632. In this controversial new volume, the author explores a new science of energy pricing and how it can be done in a way that is sustainable for the world's economy and environment. *NOW AVAILABLE!*

An Introduction to Petroleum Technology, Economics, and Politics, by James Speight, ISBN 9781118012994. The perfect primer for anyone wishing to learn about the petroleum industry, for the layperson or the engineer. *NOW AVAILABLE!*

Ethics in Engineering, by James Speight and Russell Foote, ISBN 9780470626023. Covers the most thought-provoking ethical questions in engineering. *NOW AVAILABLE!*

Formulas and Calculations for Drilling Engineers, by Robello Samuel, ISBN 9780470625996. The most comprehensive coverage of solutions for daily drilling problems ever published. *NOW AVAILABLE!*

Emergency Response Management for Offshore Oil Spills, by Nicholas P. Cheremisinoff, PhD, and Anton Davletshin, ISBN 9780470927120. The first book to examine the Deepwater Horizon disaster and offer processes for safety and environmental protection. *NOW AVAILABLE!*

Advanced Petroleum Reservoir Simulation, by M.R. Islam, S.H. Mousavizadegan, Shabbir Mustafiz, and Jamal H. Abou-Kassem, ISBN 9780470625811. The state of the art in petroleum reservoir simulation. *NOW AVAILABLE!*

Energy Storage: A New Approach, by Ralph Zito, ISBN 9780470625910. Exploring the potential of reversible concentrations cells, the author of this groundbreaking volume reveals new technologies to solve the global crisis of energy storage. *NOW AVAILABLE!*